T0325179

68th Conference
on Glass Problems

68th Conference on Glass Problems

A Collection of Papers Presented at the 68th Conference on Glass Problems

Edited by
Charles H. Drummond, III

A John Wiley & Sons, Inc., Publication

Copyright © 2008 by The American Ceramic Society. All rights reserved.

Published by John Wiley & Sons, Inc., Hoboken, New Jersey.
Published simultaneously in Canada.

No part of this publication may be reproduced, stored in a retrieval system, or transmitted in any form or by any means, electronic, mechanical, photocopying, recording, scanning, or otherwise, except as permitted under Section 107 or 108 of the 1976 United States Copyright Act, without either the prior written permission of the Publisher, or authorization through payment of the appropriate per-copy fee to the Copyright Clearance Center, Inc., 222 Rosewood Drive, Danvers, MA 01923, (978) 750-8400, fax (978) 750-4470, or on the web at www.copyright.com. Requests to the Publisher for permission should be addressed to the Permissions Department, John Wiley & Sons, Inc., 111 River Street, Hoboken, NJ 07030, (201) 748-6011, fax (201) 748-6008, or online at http://www.wiley.com/go/permission.

Limit of Liability/Disclaimer of Warranty: While the publisher and author have used their best efforts in preparing this book, they make no representations or warranties with respect to the accuracy or completeness of the contents of this book and specifically disclaim any implied warranties of merchantability or fitness for a particular purpose. No warranty may be created or extended by sales representatives or written sales materials. The advice and strategies contained herein may not be suitable for your situation. You should consult with a professional where appropriate. Neither the publisher nor author shall be liable for any loss of profit or any other commercial damages, including but not limited to special, incidental, consequential, or other damages.

For general information on our other products and services or for technical support, please contact our Customer Care Department within the United States at (800) 762-2974, outside the United States at (317) 572-3993 or fax (317) 572-4002.

Wiley publishes in a variety of print and electronic formats and by print-on-demand. Some material included with standard print versions of this book may not be included in e-books or in print-on-demand. If this book refers to media such as a CD or DVD that is not included in the version you purchased, you may download this material at http://booksupport.wiley.com. For more information about Wiley products, visit www.wiley.com.

Library of Congress Cataloging-in-Publication Data is available.

ISBN 978-0-470-34491-0
ISBN 978-0-470-37927-1 (special edition)

10 9 8 7 6 5 4 3 2 1

Contents

Foreword

The 68th Conference on Glass Problems was sponsored by the Departments of Materials Science and Engineering at The Ohio State University. The director of the conference was Charles H. Drummond, III, Associate Professor, Department of Materials Science and Engineering, The Ohio State University.

Dean William A. Baeslack, College of Engineering, The Ohio State University, gave the welcoming address. Rudolph Buchheit, Chair, Department of Materials Science and Engineering, The Ohio State University, gave the Departmental welcome.

The themes and chairs of the five, half-day sessions were as follows:

GLASS MANUFACTURING
Terry Berg, CertainTeed, Athens GA, Larry McCloskey, Toledo Engineering, Toledo OH and Gerald DiGiampaolo, PPG Industries, Pittsburgh PA

NEW DEVELOPMENTS
Carsten Weinhold, Schott, Dyrea PA, Elmer Sperry, Libbey Glass, Toledo OH and Martin H. Goller, Corning, Corning NY

GLASS MELTERS
Phil Ross, Glass Industry Consulting, Laguna Niguel CA, Tom Dankert, O-I, Toledo OH and Ruud Beerkens, TNO Glass Technology – Glass Group, Eindhoven, The Netherlands

COMBUSTION/REFRACTORIES
John Tracey, North American Refractories, Cincinnati OH, Dick Bennett, Johns Manville, Littleton CO and Sho Kobayashi, Praxair, Danbury CT

Preface

In the tradition of previous conferences, which began in 1934 at the University of Illinois, the papers presented at the 68th Annual Conference on Glass Problems have been collected and published in this proceedings as the 2007 edition of The Collected Papers.

The manuscripts are reproduced as furnished by the authors, but were reviewed prior to presentation by the respective session chairs. Their assistance is greatly appreciated. C. H. Drummond did minor editing with further formatting by The American Ceramic Society. The Ohio State University is not responsible for the statements and opinions expressed in this publication.

CHARLES H. DRUMMOND, III
Columbus, OH
November 2007

Acknowledgments

It is a pleasure to acknowledge the assistance and advice provided by the members of Program Advisory Committee in reviewing the presentations and the planning of the program:

Ruud G. C. Beerkens—TNO-TPD

Dick Bennett—Johns Manville

Terry Berg—CertainTeed

Tom Dankert—O-I

Gerald DiGiampaolo—PPG Industries

Martin H. Goller—Corning

H. "Sho" Kobayashi—Praxair

Larry McCloskey—Toledo Engineering

C. Philip Ross—Glass Industry Consulting

Elmer Sperry—Libbey Glass

John Tracey—North American Refractories

Carsten Weinhold—Schott

BATCH PLANT DUST COLLECTION — AN ENGINEERED APPROACH TO DUST REDUCTION

Don Feuerstein and Barney Olson
Middough Inc., Cleveland OH

Natalie Gaydos
PPG Industries, Pittsburgh PA

ABSTRACT

In an effort to continuously improve its processes, PPG Industries is interested in reducing dust in their glass batching operations. With this goal in mind, PPG Industries, Inc. and Middough Inc. collaborated to evaluate a batch plant's dust collection systems. The review consisted of an assessment of the existing conditions, the system's design intents, plant operations and maintenance standards. This paper will define the project approach used and the successes achieved in improving the plant systems.

CONTROL OF AIRBORNE CONCENTRATIONS

The three traditional approaches to controlling emissions and employee exposures are administrative controls, personal protective equipment and engineering controls. Administrative controls are a way of reducing the duration, frequency and severity of exposure. Examples would include job rotation and reducing the amount of time a person spends in an area. Personal protective equipment may be used to reduce employee exposure while other controls are being implemented or are not feasible or effective. Engineering controls and maintenance of those systems is another method for controlling airborne concentrations of dust and is the focus of this discussion.

PROJECT APPROACH

PPG has a number of glass facilities that utilize dry granular materials in large volumes. It was decided to evaluate the dust collection systems at these facilities to determine if engineering controls could be utilized to improve the ambient conditions by lowering the level of air particulates.

An initial site was identified that is representative in age, condition and production capacity to be evaluated. Once the site selection was completed, a project plan was formulated to monitor the progress.

The key milestones to the plan were:
1. Gather and Review the Existing Documentation
2. Site Visit / System Evaluation
3. Interview Key Stakeholders
4. Measure the Existing Conditions
5. Recommend Changes
6. Measure the Improvements / Report the Results

1. Gather and Review the Existing Documentation:

The selected site has two side-by-side glass float lines that operate 24 hours per day, 7 days per week, and an approximately 14-year rebuild cycle. The associated batch plant has six concrete stave silos located over two parallel batch collecting conveyors. An unloading shed with two rail sidings runs parallel to the silos. Stations for each raw material allow unloading by either bulk truck or railcar, with most deliveries by truck. Two silos are divided, so that there are eight unloading stations in all, four per siding. Each unloading station typically consists of an unloading hopper, vibratory feeder, bucket elevator, and transfer chute to the silo. Each station also has a dust collection system including a dust collector, blower, rotary discharge, and ductwork.

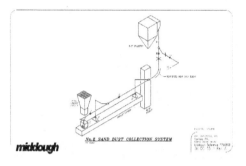

Each silo (or silo section) has two discharge/weigh systems, one over each batch conveyor. Each of these sixteen systems consists of a vibratory feeder, weigh hopper, and a second vibratory feeder that unloads the weighed ingredients onto the batch collecting conveyor. As an exception to this, the two Salt Cake systems use a double feed and weigh concept to improve measuring accuracies. The Gallery Dust Collector is located in the east end of the Unloading Shed. It serves the unloading and weighing of raw materials from all silos onto either collecting conveyor.

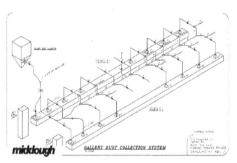

There is a Mix Tower at the east end of the silo row that houses two mix systems. Each collecting conveyor dumps to a bucket elevator, which in turn takes batch components up to a totalizer hopper to stage and check-weigh the batch into the mixer. A manual bag-fed volumetric feeder is set up to add minor ingredients to the totalizer hopper for specific recipes. The 5th & 6th Floor Dust Collector also serves to control dust from additional sources in the batch tower.

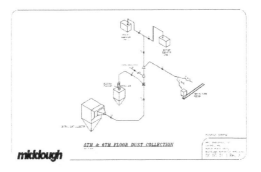

6TH & 6TH FLOOR DUST COLLECTION

middough

There are two transfer belt conveyors that take batch from the mixers to the tank feed hoppers in the main building by way of an overhead enclosed bridge. Normally Batch Line No.1 feeds Tank No.1, and Batch Line No.2 feeds Tank No.2, but a diverter valve in the unloading chute under each mixer allows batch to be fed to the alternate belt conveyor and tank. Tank No.1 Feed End Dust Collector serves the belt transfer / unload operation in the bridge, and the tank feed operation. Tank No. 2 Feed End Dust Collector serves the same function for Tank No. 2.

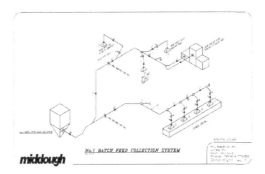

No.1 BATCH FEED COLLECTION SYSTEM

middough

The plant provided data including tables for each dust collection system with rough schematics, expected flows, measured flows, and a drawing set (plans & elevations) for the batch plant.

While these drawings provided a good background of the process, there was very little documentation for the original design of the dust collection systems. The data available did not establish the original design operating point for any of the dust collection systems such as flow rates, static pressure across the blower and duct velocity. It was assumed that the collectors and blowers were installed with the original plant, and their nameplates generally have been lost and are no longer available. Some target operating criteria has been developed by the plant for the purpose of monitoring system performance, but the missing data made the evaluation more challenging.

2. Site Visit / System Evaluation:

The process operations appeared to be reasonable for the plant requirements. The batch plant is typical for this type of operation. Capacity appears to be well matched for operations, and the original equipment selection appears to be logical and functional. The material transport systems were basically industry standard and reasonable for the system size. In general, all points of material transfer had dust collection pick up points.

The original sizing of the batch dust collection systems could be improved in terms of the bag cloth area. This may not have been a problem initially, but as new bags became slightly impinged, flows dropped below conveying velocities, and the systems all became difficult to maintain.

A review of the dust collection equipment (the blowers and the dust collectors) resulted in a decision that the existing equipment, even at thirty plus years old, was in reasonably good condition. Therefore, an effort was made to retain the existing equipment. It was determined that the original design of all twelve (12) of the batch dust collection systems provided ample air flow volume, but minimal dust conveying line velocities. This situation caused chronic maintenance and operational issues and compromised the plant's effort to maintain proper housecleaning. Therefore, the focus of the engineering solution was aimed at the duct system, since the dust collection equipment was determined to be reasonable for the application.

3. Interview Key Stakeholders:

Key personnel in management, engineering, and operations were interviewed to gain a working perspective of concerns with operation and maintenance of the systems. The discussions were centered around the adequacy of the existing process and dust collection equipment, especially in regards to dust management.

There was a general consensus that all of the dust collection systems are hard to maintain and are prone to plugging. It is difficult to keep the bags from fouling in some of the systems as well. There were no major issues raised regarding the process equipment.

4. Measure the Existing Conditions:

The remainder of this paper concentrates on the analysis of three of the Truck / Railcar Unloading Shed Dust Collection Systems. While some of the other batch dust collection systems have a larger distribution network, their problems, analysis, and potential solutions are very similar. The eight truck/railcar Unload Dust Collection systems were all installed with the original plant. They are basically identical with the exception that some of the ductwork has been individually modified for various reasons. They each utilize a reverse pulse jet dust collector with 236 ft^2 of polyester fabric bags. All have pick-ups at both the feed and discharge ends of the vibratory conveyor, and a dust recycle chute from the collector hopper back to the conveyor. The soda ash and aragonite pick-ups were by-passed with a short pick-up straight to the bucket elevator due to chronic plugging, but these new legs plug occasionally. Ducts were added to both of the sand systems to draw in room air, and to provide a common crossover.

EXISTING SYSTEMS:

Rail / Truck Unload Shed - Typical System #1

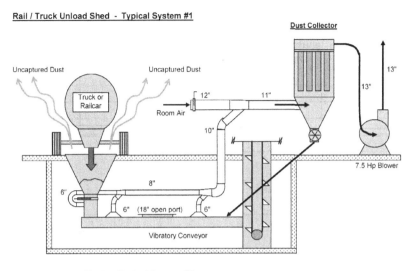

Rail / Truck Unload Shed - Typical System #2

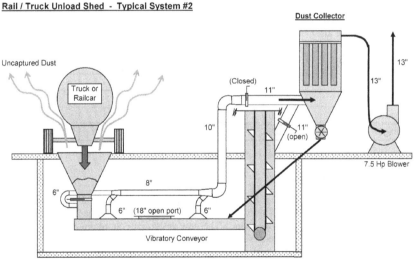

The following tables show a summary of the plant flow measurements:

Table I. Rail / Truck Unloading Shed - Measured Dust Collector Flow Rates

Plant Data taken in January 2007

Raw Material	Flow Rate (cfm)	Main Header Size	Velocity (fpm)	Air to Cloth Ratio
#1 Sand (low Iron)	864	11"	1,370	3.7
#2 Sand	873	11"	1,390	3.7
Soda Ash	plugged	11"	0	0.0
Average Flow	579		920	2.5

Plant Data taken in June 2007

Raw Material	Flow Rate (cfm)	Main Header Size	Velocity (fpm)	Air to Cloth Ratio
#1 Sand (low Iron)	1,900	11"	3,020	8.1
#2 Sand	1,530	11"	2,430	6.5
Soda Ash	1,680	11"	2,670	7.1
Average Flow	1,700		2,700	7.2

Target Plant Data after Modifications

Raw Material	Flow Rate (cfm)	Main Header Size	Velocity (fpm)	Air to Cloth Ratio
#1 Sand (low Iron)	1,420	8"	4,070	6.0
#2 Sand	1,420	8"	4,070	6.0
Soda Ash	1,420	8"	4,070	6.0
Average Flow	1,420		4,070	6.0

5. Recommended Changes:

Dust system line velocities should always be greater than the saltation velocity for the particular dust being handled. This will vary for any given material based on particle size density (PSD), moisture and particle shape. The new design criteria minimum line velocity for free flowing non-hazardous powders is set at 3,500 feet per minute.

Reverse pulse jet dust collectors can operate across a range of flows, but differential pressure across the fabric increases as flow rate increases. At some point, airflow becomes too great for the bags to effectively clean. This maximum airflow depends on collector design, pulse air pressure and material characteristics. The new design criteria for the maximum air to cloth ratio is set at 6 to 1. This assumes the powder load is light (dust collection, not a product collector) and that the dust collector is of a reasonably good design with adequate interstitial area. The air to cloth ratio uses air flow and the cloth area (sum of all bags), so the net value is actually the face velocity of air across the fabric in feet per minute. The velocities in the Unloading Shed Dust Collection Systems are too low to provide adequate dust collection.

Specifically, if any of these systems were started clean and fresh after a complete overhaul (new bags, clean ducts, repairs), duct velocity would initially be adequate for conveying, but the dust collector would not be capable of adequately cleaning itself. The dust that was dislodged by the pulsing action would immediately be re-entrained and carried back to the bag surface. As powder builds up on the bags, velocity would gradually drop to an equilibrium value where the bags clean, but at a resulting duct velocity that is too low to properly convey powder, so the ducts continue to plug. Soda ash and aragonite are additionally troublesome as the settled hygroscopic powder in the ducts and on the bags would readily absorb moisture, become tacky, and attract yet more powder out of the air stream.

The goal of most dust collection systems is simply to create a slightly negative pressure at all points inside the dust-bearing equipment of a process or conveying system so that any entrained dust cannot escape at potential leak points or design openings in that equipment. Too little suction results in dust leaks, so many designers often provide extra capacity. Too much suction can create a bigger problem, however. In addition to the waste of capital in an oversized dust system, excessive suction can induce airflows inside the process that can result in product entrainment and product loss at transfer points. In some cases, fine minors or micro ingredients can be selectively 'vacuum cleaned' right out of a system, resulting in inaccurate batch formulations in addition to excessive collection of powders as waste material.

The design criteria was to provide a flow equal to 5 times the sum of the volume of displacement air from solids transfer and the expected or measured leak rate of the system seals, plus 150 fpm inflow at any openings to the system (such as bag dump stations or open inspection hatches). Each of the truck unload systems require approximately 150 cfm for material displacement below the dump hopper, 50 cfm for leaks, and another 30 cfm for the open face area of an open 6" round inspection port in the vibrating conveyor. Each system as originally installed provides several times this requirement, even at low velocity.

Since the powder unloading from the truck or railcar effectively forms a seal leg in the dump hopper, another 150 cfm is required for displacement and 450 cfm for the open face area to account for the new partial covers provided to reduce the open grating area to less than 3 ft^2.

RECOMMENDED SYSTEMS:

Rail / Truck Unload Shed - Typical System #1

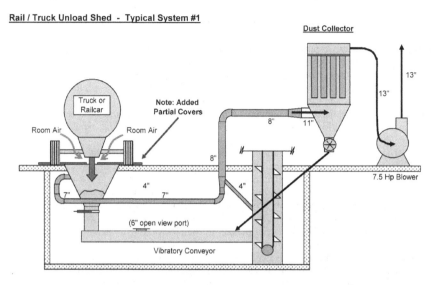

Rail / Truck Unload Shed - Typical System #2

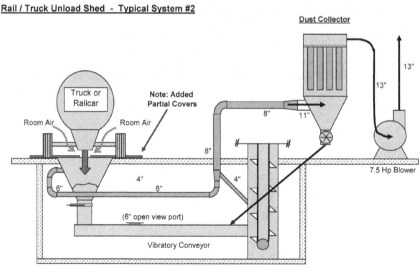

Our analysis of these systems determined the existing collector / blower combinations provide more than adequate flow for all re-sized pick-up points after adjusting to a 6.0 air to cloth ratio operating point. Most of this capacity will now be directed to 2 new pick-ups at the top of the unloading hopper, while a single 4" pick-up on the bucket elevator will provide adequate suction to all equipment downstream of the dump hopper seal leg. The main header to each collector was also reduced from 11" to 8" to establish conveying velocity. Flow rate and velocity at each pick-up and trunk is balanced as follows:

Table II. Rail / Truck Unload Shed
Target branch flows after balancing

Target Branch Flows - Unloading Systems

Duct / Pick-up	Flow Rate (cfm)	Velocity (fpm)
6" from hopper	750	3,820
4" from hopper	350	4,010
6" trunk	1100	5,600
4" from elevator	350	4,010
8" trunk (total flow)	1450	4,150
13" to blower (clean)	1450	1,700

It was recommended that Gore-Tex be specified as the bag material for the two hygroscopic powders, aragonite and soda ash. This PTFE membrane-lined filter fabric provides a relatively inexpensive additional safeguard against caking of material on the bags

6. Measure the Improvements / Report the Results:
Flow and pressure drop data will be taken on all three systems 1 week, 1 month, and 3 months after the systems are back on line. We expect to see consistent performance that will demonstrate that the modified systems are truly low maintenance as compared to the original installation.
Unload shed area measurements will also be measured to demonstrate the environment improvements.

FOREHEARTH COLORING ADVANCEMENTS

John M. Bauer
Ferro Corporation
Orrville, OH

ABSTRACT

The forehearth coloring process has been assisting glass manufacturers with the production of custom glass colors since the 1970's. The forehearth coloring process was started on forehearths pulling 20 tons per day for the production of cosmetic containers. Color forehearth design has gone through many changes to keep pace with the advancements in the glass manufacturing process. Over the past 30 years, the glass manufacturing process has advanced with the introduction of larger forming machines and wider and deeper forehearth channels. The forehearth coloring process has kept pace with these manufacturing changes and continues to serve the needs of all glass manufacturers.

This paper will review the historical design of color forehearths and discuss the advancements that have been made over the past several years. These advancements have allowed Ferro to increase the coloring capacity of color forehearth installations to a level of 150 tons per day. This paper will detail the general layout and design of a color forehearth. The stirring zone configuration, a critical point for the success of a color forehearth, will be reviewed for different forehearth capacities. The auxiliary equipment required for the conversion of a color forehearth will be briefly examined.

HISTORY OF FOREHEARTH COLORING

The addition of color to glass can be traced to the earliest of times. Historical glass pieces from Egyptian and Phoenician cultures can be viewed with definitive color streaks and overall coloring of the glass articles. The forehearth coloring process can only trace its history to the 1970's when Ferro and several glass manufacturers started experimenting with different methods color could be produced on a furnace without the furnace completing a color transition. While a color transition is not difficult to complete, the production downtime spent going from one color to another and the limited production associated with each new color made a furnace transition for short color runs very impractical. Glass manufacturers were looking for a method to change color quickly on a single forming line, produce the limited number of items in the new color and return the machine production to the base glass from the furnace without affecting the other forming machines. The forehearth coloring process was born.

The driving force behind the forehearth coloring process has changed over the years. The initial driving force was from the cosmetic industry. Cosmetic manufacturers were looking for a wide range of colors in limited quantities while the glass manufacturers were looking for a method to quickly and efficiently produce these short runs of specialty colors without upsetting the overall furnace production. Tableware manufacturers soon followed in the steps of the cosmetic manufactures with the production of new and varied tableware colors to match the current color trends. Container manufacturers soon learned that a color forehearth could be used to produce their limited quantities of emerald and Georgia green glass. Container manufacturers continue to produce emerald and Georgia green glass but are also using the color forehearths to produce containers for the wine and spirits manufacturers. Common colors now include cobalt blue, smoke brown and antique green from a deadleaf green glass base. Developing countries and economies around the world are now showing their demand for colored glass in all of the glass manufacturing areas from containers to tableware, cosmetics and architectural blocks and flat glass.

EARLY FOREHEARTH COLORING PROCESSES

It is very important to acknowledge the early work that was completed on the development of the forhearth coloring process as most of this information is still practical with current color forehearths. The initial development work focused mainly on the stirring zone of the color forehearth and the methods to get the color efficiently mixed into the base glass as the glass moved through forehearth.

The mixing process used today was actually developed on a scale model of a color forehearth in the 1970's. A one quarter scale Plexiglas channel containing a 600 poise silicone fluid was built to model the behavior of the glass flowing through a color forehearth channel. The model forehearth proved to be a most valuable tool in the development of the stirring zone.

Model Forehearth Showing Stirring Studies with Heavy Oils

The testing completed on the scale model forehearth identified the key areas of the mixing process. The researchers were able to determine that auger-type blenders were the most satisfactory for the blending and mixing of the color into the glass. The auger blender was chosen for a variety of reasons including their initial cost, ease of installation, durability, space requirements and their overall commercial availability.

The forehearth model also allowed the researchers to determine the optimum dimensions of the blenders, the spacing of the blenders within a single bank, the spacing between the stirring banks and the most favorable direction of stirrer rotation. The best results were obtained when the blenders had sufficient pitch-length to break the surface of the glass and were operated so that all of stirrers were rotated with a lifting thrust. What may have been one of the most important discoveries was that the pull on the forehearth does not affect the spacing of the blenders. This last discovery allowed for a standardization of the stirring zone on all color forehearths.

Additional recommendations gleaned from the scale model studies showed that the stirrers within an individual stirring bank should all be operated in the same direction and that the direction of rotation should be alternated from bank to bank. This operation has the effect of extending the dwell time of the glass in the stirring section. One of the critical dimensions established from the model studies was that the outside stirrer should not be more than 2½ inches from the channel sidewall toward which it is pulling the glass.

FOREHEARTH COLORING BASICS

There are two critical parameters that must be met by each color forehearth. The first basic parameter is that the color forehearth must operate at the proper time and temperature relationship that provides for complete melting of the colorant material. Secondly, the stirring zone must blend the color into the base glass to produce a homogeneous color at the end of the stirring zone when all blending of the glass ceases.   All color forehearths are based on the same general layout. Every color forehearth is broken down into three zones: Preheat Zone, Melt Zone and Stirring Zone.

Preheat Zone

The preheat zone is the first zone of a color forehearth located from the entrance of forehearth to the centerline of the feed tube. The preheat zone is used to maintain or increase the temperature of the glass prior to the melt zone. The length of the preheat zone can vary from a minimum length of 12 inches to several feet. The primary factor that determines the length of the preheat zone is the temperature of the incoming glass. Higher incoming glass temperatures (1250°C - 1270°C) mean that the preheat zone can be shortened since it will be used solely to maintain the temperature of the glass before the colorant addition. If the incoming glass has a low temperature, then the preheat zone will need to be extended and auxiliary combustion employed in the zone to raise the temperature of the glass. Additional combustion can be completed with a second row of forhearth burners or auxiliary burners placed in the crown.

Melt Zone

The melt zone is the middle or second zone on a color forehearth. The melt zone length is measured from the centerline of the feed tube to the centerline of the first stirring bank. The feed tube is installed through the cover block of the forehearth allowing for the transfer of the colorant material from the outside of the forehearth to the inside of the forehearth.

The color concentrate introduced into the feed tube falls onto the surface of the glass where the melting of the colorant started. The melt zone is sized to insure that adequate time is provided, even at the maximum operating tonnage of the forehearth, to allow the color concentrate to completely melt and degas before any mechanical mixing of the color concentrate is started in the stirring zone.

Attempting to mix the color into the glass before all degassing is complete could lead to the entrainment of the melting colorant into the body of the glass which would introduce blisters into the glass. The time and temperature conditions remaining in the forhearth would not be enough to eliminate these blisters from the body of glass.

Stirring Zone

The stirring zone is the last zone of the color forehearth. The stirring zone is measured from the centerline of the first stirring bank to the centerline of the last stirring bank. In this zone, a series of refractory stirring banks blend the color into the base glass to produce a homogeneous color before the glass exits the zone after the last stirring bank.

The number of stirrers in each stirring bank depends on the channel width. A wider the channel will require a greater number of stirrers cover the larger cross sectional area. The auger stirrers in a color forehearth typically have a flight diameter between 6½ inches and 7¼ inches. The flight length of a stirrer should be equal to the glass depth. A flight length equal to the glass depth insures that the stirrer flight will just break the surface of the glass when the stirring bank is properly installed 1 inch above the channel bottom.

The physical spacing of the stirrers in each stirring bank is consistent in all color forehearth designs. The distance between neighboring stirrer flights is set at a distance of 1 inch. As we learned from the scale model development, the separation between the outside stirrer flights and the channel sidewall should never be more than 2½ inches.

A color forehearth will use a minimum of three stirring banks with the maximum number of banks determined by the tonnage being pulled through the stirring zone. Color forehearths typically have 3-5 sets of stirring banks. All of the stirrers in a single stirring bank are rotated in the same direction. All of the stirrers on a stirring bank will either be all right hand or all left hand stirrers. This insures a consistent movement of the glass by all of the stirrers in a single stirring bank. The general recommendation is that each stirring bank pumps the glass in an upward direction. This means a stirring bank containing right-hand stirrers will operate in a clockwise motion. Left-hand stirrers will turn in a counter-clockwise rotation to achieve the upward pumping action.

Looking at the stirring zone as a whole, neighboring stirring banks will operate in alternating directions. If the first bank of stirrers is using right-hand stirrers and turning in a clockwise direction pumping the glass in an upward direction, the second bank of stirrers will contain left-hand stirrers and the bank will be operated in a counter-clockwise direction to pump the glass in an upward direction. The third stirring bank will be the same as the first bank and the variance of the stirring direction will continue through the entire stirring zone.

EARLY COLOR FOREHEARTHS

Now that we have an understanding on the basic operation and configuration of a color forehearth, let's look at some of the early color forehearth layouts. The early color forehearth layouts were designed to pull at a tonnage of 20 tons per day on channels that had a 26-inch width and a 6-inch glass depth. Here are some of the details on this color forehearth.

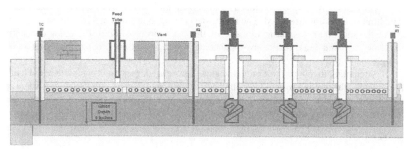

Table I. Color Forehearth Dimensions for a 20 Ton Per Day Color Forehearth

Channel Width	26 inches
Minimum Glass Depth	6 Inches
Preheat Zone	24 Inches
Melt Zone Length	56 Inches
Melt Zone Residence Time	28.2 Minutes
Stirring Zone Length	52 Inches
Stirring Residence Time	33.3 Minutes

The coloring capacity of these color forehearths was very limited due mainly to their physical construction. The channel width of 26 inches and the glass depth of 6 inches were limiting factors that could not be overcome. The only way to increase the capacity of the color forehearth would be to increase the length of the melt zone and install additional stirring banks. The physical length of the color forehearth was typically not long enough to accommodate these longer zones. Changing the glass depth by changing the channel blocks or designing a new foot print for the whole forehearth was too costly so most of the glass manufacturers would work within the existing forehearth layout.

EARLY COLOR FOREHEARTH ADVANCEMENTS

The successful introduction of the forehearth coloring process in the cosmetics area led other glass manufacturers to adopt the forehearth coloring process to their glass manufacturing applications. Tableware manufacturers were interested in color as a natural way to increase the size of their product offerings while container manufacturers were interested in forehearth coloring for the production of emerald green and Georgia green. Additionally, the globalization of glass production, which continues to this day, increased the number of color forehearths.

The shift of the color production from the lower pulled cosmetic color forehearths to the higher pulled tableware and container color forehearths forced the first redesign of the standard color forehearth layout. Cosmetic color forehearth channels that were just 26 inches in width were giving way to container forehearth channels that now had a width of 36 inches.

Two additional areas had to change on the container color forehearth to allow for the installation of a color forehearth. The first area that was addressed was the glass depth. Most container forehearths were operating with glass depths of 6 inches or less to facilitate the temperature conditioning of the glass. The shallower glass depth and higher tonnages worked against each other shortening the residence time for the melting and stirring functions. Secondly, most container forehearths were now using upper structure cooling on the forehearth with special roof blocks. The roof blocks made it impossible to install a feed tube opening or the openings required for the stirring banks. The color forehearth would have to find a new location.

The combination of the forehearth glass depth and the upper structure cooling zones pushed the color forehearth location from the forehearth rearward into the alcove or the working end section. This turned out to be a good move for the forehearth coloring process. The working end offered some advantages that fit nicely into the standard color forehearth design. Most working ends were operating with wider channels but they also had a deeper glass depth and additional length that was not available on the forehearths. The entrance to the working end was closer to the furnace throat which meant that the glass entering the color forehearth was coming in at a higher temperature and the need for a longer preheat zone was decreased. Most working ends did not have any type of upper structure cooling so the openings for the feed tube and the stirring banks could be accommodated with minimal refractory changes.

These early container color forehearths were typically pulling at a rate of 75 tons per day, a four-fold increase over the early color forehearth tonnages but still low when compared to today's standards. Here are some typical dimensions from an early container color forehearth.

Table II. Color Forehearth Dimensions for a 75 Ton Per Day Color Forehearth

Channel Width	36 inches
Minimum Glass Depth	10 Inches
Preheat Zone	24 Inches
Melt Zone Length	92 Inches
Melt Zone Residence Time	28.5 Minutes
Stirring Zone Length	78 Inches
Stirring Residence Time	28.5 Minutes

While the side profile looked about the same, the cross sectional view of the stirring zone was undergoing a definite change of appearance. The stirring zone on a 26-inch channel would use three stirrers in each stirring bank. The 36-inch channel would need additional stirrers to cover the cross sectional area while maintaining the clearances that were developed in the model studies. A different stirring configuration was required.

The 26-inch channels were using three stirrers with 6½-inch diameter flights which provided good coverage of the channel cross section. The flights were separated by a distance of 1 inch and the distance between the outside stirrer and the channel sidewall was 2¼ inches.

Using four of the same stirrers on the new 36-inch channels wound not provide the same coverage. The stirrer flights could be set at a distance of 1 inch but this would leave 3½ inches between the outside flights and the channel sidewall. The early studies showed that this outside distance should not be larger than 2½ inches or the mixing efficiency would be lowered. Adding a stirrer was not feasible as the stirrers would be too close to each other and to the channel sidewalls. To close the gap, larger stirrers were designed and installed. The new stirrers had a 7¼-inch flight diameter. The new stirring setup kept the 1 inch separation between the stirrer flights and decreased the distance between the outside stirrer and the channel sidewall to 2 inches.

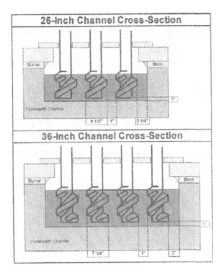

RECENT COLOR FOREHEARTH ADVANCEMENTS

Glass container manufacturers were not finished with the changes to the manufacturing process. Forming machines continued to increase in size and tonnage which meant the forehearth channels had to be widened to keep pace with the higher tonnages. In the early part of this century, glass manufacturers started inquiring about color forehearths for a channel width of 48 inches and tonnages over 110 tons per day. The restrictions on the placement of the color forehearth in the forehearth section still existed so the working end was naturally targeted for the installation of the color forehearth.

The basic color forehearth design principles that were followed when the process was converted from the 26-inch to 36-inch channels were targeted for the 48-inch channels. The longitudinal view of the color forehearth remains unchanged except for the zone lengths but the cross sectional view through the stirring bank shows a major change. The stirring zone now shows a 48-inch channel cross sectional area but the design principles of the stirring zone need to remain the same. These principles dictate a 1-inch separation between the stirring flights and a maximum separation of 2½ inches between the outside stirrers and the channel sidewall.

The new stirring design for the 48-inch channel consists of 6 stirrers. The stirrer flights will go back to the 6½-inch flight diameter that was used on the original color forehearths. The distance between the stirrer flights will continue at the 1-inch separation and the separation between the channel sidewall and the outside stirrers is set at 2 inches. Most of the color forehearths that have been installed on the 48-inch channels use a glass depth of 12 inches. The deeper glass depth increases the residence time for the glass allowing the color forehearth to complete the melting and the stirring functions within the color forehearth spacing.

Table III. Color Forehearth Dimensions for a 120 Ton Per Day Color Forehearth

Channel Width	48 inches
Minimum Glass Depth	12 Inches
Preheat Zone	48 Inches
Melt Zone Length	92 Inches
Melt Zone Residence Time	28.5 Minutes
Stirring Zone Length	78 Inches
Stirring Residence Time	28.5 Minutes

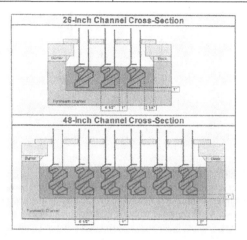

The color forehearth described in the preceding table uses 4-banks of 6-stirrers to homogenize the glass in the stirring zone with maximum tonnages approaching 120 tons per day.

In early 2005, Ferro assisted with the startup of the largest domestic color forehearth to date. The color forehearth was installed on a 48-inch channel that was operating at a 12-inch glass depth. The stirring zone had a stirring configuration consisting of 5-banks of 6-stirrers. The maximum rated capacity of the color forehearth was set at 140 tons per day but it was our belief that the tonnage for lighter, low feed rate colors could have easily topped 160 tons per day.

The largest coloring section presently operating is based on a 51-inch channel width. The channel uses 6 stirrers that have a 7-inch flight diameter. The standard separation in the stirring zone was maintained on this color forehearth design as the stirrer separation was set at 1-inch and the outside separation was 2 inches.

ADDITIONAL COLOR FOREHEARTH ADVANCEMENTS

The color forehearth advances are not limited to the color forehearth design. The auxiliary equipment required to weigh and deliver the color concentrate into the color forhearth has undergone several changes over the years. The volumetric feed auger that was used to deliver the colorants on the original color forehearths has been replaced with an electronic weigh scale. The feed rate variations that were inherent with the volumetric feeding of different particle sizes are negated with the weighing of the colorant materials. The initial version of the weigh scale improved on the feed rate accuracy but the current electronic weigh scale took the accuracy to another level. Setting the target weight, the vibratory speed and the electronic timer produces accurate product feed rates. Variations of less than +/- 1 gram over days of production are not uncommon even when weighing over 400 grams per cycle.

Improved colorant formulations now allow for the production of a wide range of colors. The color range is practically endless but we need to remember that a color forehearth cannot take an oxidized glass to a reduced glass for obvious reasons so a reduced amber glass is still not possible from an oxidized flint glass base. Recent advancements do allow for the coloring of reduced type glasses. Deadleaf green glass is now being regularly converted to both antique green and French green on color forehearths serving the expanding worldwide wine industry.

CONCLUSION

The forehearth coloring process that was started in the 1970's has gone through several changes with respect to the layout of the color forehearth zones and the coloring equipment to adjust to the increasing pull and coloring demands of glass manufacturers. The forehearth coloring process will continue to make improvements to the process to offer glass manufacturers the flexibility to deliver the properly colored glass to their customers.

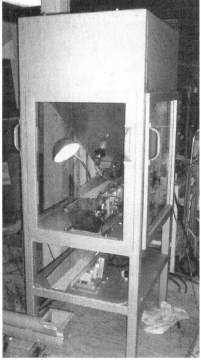

Weigh Scale Control Panel and Weigh Scale Unit

Weigh Scale and Delivery Tube Conveyor on a Color Forehearth

Stirring Bank Configuration

Operating Stirring Banks Under Cover Blocks

GOB TEMPERATURE CONTROL

Brent Illingworth and Jim Knope
BASF Catalysts LLC
Portland, Oregon, USA

Steve Nixon and Alan White
Advanced Control Solutions, Inc.
Sylvania, Ohio, USA

Julian Railey
Saint-Gobain Containers, Inc.
Muncie, Indiana, USA

ABSTRACT

Significant quality and throughput improvement to the glass container manufacturing process has been demonstrated by a new control method that directly controls the gob temperature. The new temperature control method combines ACSI's predictive-adaptive model based control system with BASF's Exactus® advanced high-speed optical gob temperature sensor. The gob temperature sensor provides an accurate and repeatable measurement of each sheared gob. The model based control system continuously adjusts the upstream process to ensure that the gob temperature is at the setpoint regardless of environmental conditions. The system has been field tested at the Saint-Gobain Containers, Inc. – Dunkirk, Indiana, USA plant, and has demonstrated the ability to improve gob temperature stability substantially as compared to conventional PID control. Results of implementing the new control method include improved manufacturing efficiency, product quality, and reduced gob temperature recovery time following a job change.

INTRODUCTION

Conventional methods of controlling gob temperature do not include direct gob temperature measurements. The operator monitors the equalizing thermocouple measurements as a reference and adjusts individual forehearth zones in an attempt to maintain a stable glass temperature before it enters the forming process. The actual gob temperature can only be assumed or measured with a conventional optical pyrometer which is often unreliable and incapable of being used for closed-loop control.

To eliminate the uncertainty of the actual gob temperature, ACSI and BASF have developed a solution that accurately measures and controls the gob temperature. The system utilizes direct measurement of each gob through the use of the Exactus® advanced gob temperature sensor (GTS) produced by BASF. By combining the measurement from the GTS with ACSI's advanced model based control (BrainWave®), exceptional gob temperature stability and control is achieved at the point of entry to the forming process. This solution, described below, has provided significant benefits for Saint-Gobain Containers, Inc. including reduced variation in the production process and reduced gob temperature recovery time after a job change.

THE ADVANCED GOB TEMPERATURE SENSOR

Conventional optical temperature sensors typically measure light energy using a photodiode sensitive to a range of infrared wavelengths. The signal from the photodiode is amplified with hard-wired amplifiers, providing a calibrated output of target temperature. These instruments usually suffer from a poor tradeoff between repeatability, speed, resolution, and data processing capabilities.

Recent advances in microelectronics circuits allow for amplification and measurement circuitry to be brought onto a single microprocessor, allowing for direct digital measurements which are faster, more accurate, and much less prone to drift over time. Additionally, the onboard microprocessor can automatically compensate for ambient temperature changes and provide sophisticated processing of measurement data. The result is an instrument that can process up to 1000 measurements per second with greater accuracy and nearly zero instrument drift.

The benefits of accuracy and very low drift are obvious for gob, or any other process temperature measurement. The benefits of high speed are essential for gob temperature measurements. Depending on the gob length and velocity, 1000 readings per second provide 25 to 40 temperature measurements of each gob. Each of these measurements can be output digitally and/or processed by the instrument and output as a single average temperature of the gob. Conventional IR temperature sensors, with their low speed, typically have to look up at the orifice when attempting to measure the gob temperature; the upward viewing angle leaves these sensors susceptible to contamination from shear spray and other contaminants. The GTS, with its much faster measurement rate, can be aimed to measure the gobs well below the shears. This allows for a downward viewing angle so the optics are much less likely to be contaminated.

Instrument optics are designed to provide a very small, concise measurement spot with 99% of the light energy within a circle 15 mm or less at a focal distance of 2 meters. The optics are coupled to the measurement electronics with high temperature rated fiber optic cables.

The optics are contained in a protective housing with plant instrument air providing a continuous flow of air over the optics and out the sight port, ensuring the optical components stay clean. The optics housing is installed in a heavy gauge metal adjustable mount which allows for side-to-side and vertical adjustment of aim. Bright green diode lasers are utilized to provide an exact image of optics alignment and focus quality. Figure 1 displays the GTS schematically.

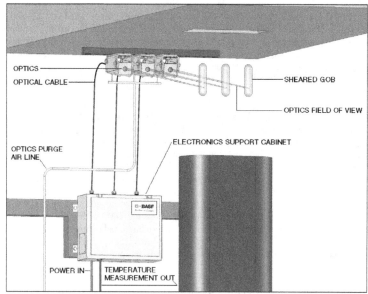

Figure 1: Schematic of gob temperature measurement system

MEASUREMENT WAVELENGTHS

Two primary techniques exist for optical temperature measurement; single wavelength (sometimes called brightness measurements) which correlates the intensity of light at one wavelength to temperature, and dual wavelength (sometimes called two-color or ratio measurement) which measures the radiant intensity at two wavelengths and correlates the ratio of those intensities to temperature. The high-speed, dual wavelength technique as used by the BASF Exactus gob temperature sensor is essential for gob temperature measurement.

The transparency of container glasses varies widely with color, temperature, and wavelength of light. The transparency influences how deep the effective temperature measurement is into the glass. In the visible and near infrared, amber and green glass are partially transparent such that the radiation emitted by the gob originates over a depth of 0 to 10mm. Flint glass is much more transparent and radiation emitted in the visible to near infrared wavelengths originates from depths of 0 to 100mm. Outside the near infrared, container glass becomes fully opaque and longer wavelength optical measurements of gobs become surface temperature measurements which will vary with environmental conditions such as ambient air and shear spray temperature.

Beneath the air-glass boundary layer, the glass is quite isothermal for a particular horizontal section of the gob. For colored container glasses such as amber and green bottle glass, the measurement wavelength must be short enough that the interior of the gob is measured. Flint glass is highly transparent at near infrared wavelengths such that the emitted radiation originates from the entire depth and will depend on the diameter of the gob. The dual wavelength technique used by the GTS solves the concern of gob diameter dependence because the glass transparency is essentially equal at the two measurement wavelengths and any variation in gob thickness will not significantly affect the ratio.

Shear spray and smoke can also affect gob temperature measurements. Each of these will attenuate the light energy from the gob that reaches the instrument optics. Single wavelength measurements are strongly affected by shear spray and smoke. Dual wavelength measurements are usually unaffected by shear spray but intense smoke can cause a few degree measurement change; typically this change is short term and is averaged out in the data acquisition hardware.

PEAK PICKING

After shearing, the gobs fall freely, accelerating as they approach the distributor. Gobs are typically measured below the drip pan where the gob velocity is usually greater than 13 ft/sec. At this speed, about 40 independent temperature measurements will be taken of a 6-inch long gob. Typical gob temperature profiles appear similar to what is shown in Figure 2.

In order to provide the gob temperature to the control system, a type of signal processing is necessary to isolate the gob temperature from the background measurement when a gob is not in the instrument's field of view. 'Peak picking' systems have been available in measurement systems for some time. When a high temperature occurs, the output will increase to a value indicating the maximum temperature measured by the instrument, commonly called the 'peak temperature'. Conventional peak picking instruments can only output the highest value measured; for gobs, the highest value measured is often not desirable. In both Figure 2 and Figure 3 the leading and trailing edge of the gob data stream show values significantly greater than is measured in the middle of the gob data. The high measurements are caused when the dual wavelength measurement views the hemisphere shaped top and bottom of the gob; these are optical effects and are not real glass temperature variations. Conventional peak picking systems will output these erroneous values causing false maximum readings to be processed as gob temperatures.

The BASF Exactus gob temperature sensor has the ability to eliminate these values from the measurement by storing numerous measurement points and processing them as a group. The user configures the measurement with two pieces of information, the minimum possible gob temperature and the number of measurement points to ignore from both the leading and trailing edge of the gob.

For gobs in a general container manufacturing operation, the minimum temperature might be 2010°F and nine data points ignored from each end of the gob measurement data stream. In the example shown in Figure 3, there are 42 valid measurements above the 2010°F-threshold temperature. Removing nine measurements from each end of the measurement profile leaves 24 valid measurements of the gob. The user has the option of transmitting the maximum or average temperature in the region of interest. The average temperature is typically used because it provides a better indication of the amount of heat transferred to the forming machine from the gobs.

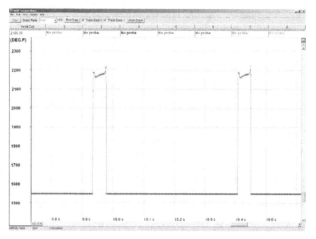

Figure 2: Gob temperature profiles

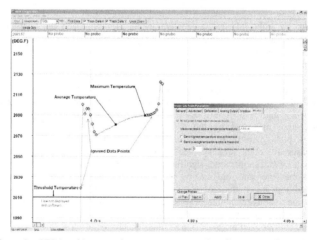

Figure 3: GTS 'peak' processing – measurement data from one gob displayed

THE SIGNIFICANCE OF THE GOB TEMPERATURE

Continuous monitoring and controlling of the gob temperature allows container manufacturers to remove thermal variation from the forming process. Because the gob temperature measurement is unaffected by the depth of glass in the forehearth as immersion thermocouples are, it is a better measurement to use when attempting to repeat process conditions.

GOB TEMPERATURE VARIATION

The GTS has monitored the gob temperature on several lines that have conventional PID control. The data has indicated that the gob temperature can fluctuate by as much as 16°F while the equalizing zone thermocouples indicate stable behavior. It has also been realized that the mechanics in the feeder have a significant impact on the gob temperature. For example, adjustments to the shearing and gobbing process have shown to change the overall gob temperature and the temperature relationship amongst the gobs from the different orifices. It has also been observed that the gob temperature commonly oscillates at the tube rotation frequency.

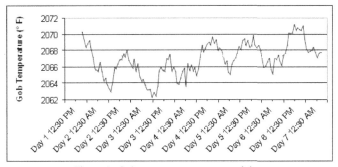

Figure 4: Gob temperature over several days

Figure 4 shows how the gob temperature can fluctuate day-to-night. While this data was acquired, the equalizing zone thermocouples indicated stable behavior.

CONTAINER DEFECTS AND GOB TEMPERATURE

The presence of container defects was compared to gob temperature on a continuous job, amber glass container line with PID control. It was observed that changes as small as a few degrees in gob temperature impacted the presence of seal surface and split finish defects. Figure 5 shows the normalized quantity of containers with either defect compared to the gob temperature. Because the forehearth was controlled with conventional PID, the gob temperature varied significantly as explained above. The gob temperatures were divided into three bins and tracked to the presence of container defects. As shown in Figure 5, an effective optimum gob temperature can be determined with the information from the GTS. The data shows that on this particular line, the quantity of containers with either defect is significantly reduced when the gob temperature is within ±2.5°F of the optimum temperature.

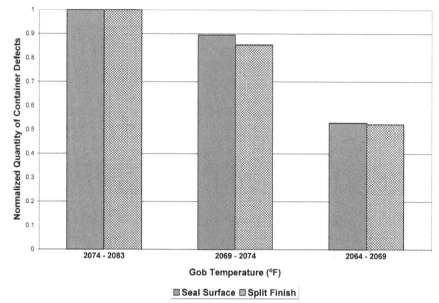

Figure 5: Container defects and gob temperature

MODEL BASED CONTROL (MODEL BASED CONTROL VERSUS PID)

Model based controllers (MBC) are outperforming PID control in glass applications and therefore have become a preferred installation by manufacturers. The MBC quickly responds to process disturbances and reacts quickly to stabilize temperature variations. A model based controller creates models for each control/process variable and feed forward input. The ideal model then anticipates changes needed to maintain consistent glass temperature. Once the optimum process is modeled, model based control

• Predicts control actions required to drive the glass temperature to setpoint quickly without overshoot.

• Continuously adapts to process and production changes automatically for better control without loop tuning.

• Models feed forward inputs and updates control actions to quickly stabilize temperature variation.

CONTROLLING THE FOREHEARTH AS A UNIT

Typically, the 9-point grid is monitored to determine temperature stability in the forehearth. The operator assumes that there is a direct relationship between gob temperature and the 9-point grid. The operator adjusts the individual zone temperatures in an attempt to achieve the desired gob temperature. As explained above, data from direct gob temperature monitoring has shown this method of control to be unreliable from both an accuracy and stability standpoint.

This new control method allows the grid temperatures to be controlled directly using the model based system. This system understands the interrelationships among zones; therefore, it removes the complications that would normally be difficult for operators to resolve. The model based controller:
- Thinks of the forehearth as a unit, not as individual zones
- Prioritizes temperature readings to determine the most important temperatures
- Allows zones to work together rather than fight each other

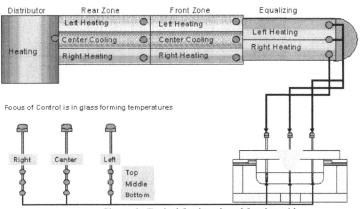

Figure 6: Typical forehearth and 9 point grid

MASS FLOW TEMPERATURE CONTROL

The mass flow temperature (MFT) is made up of an average of nine equalizer triplex temperature values. Each of these nine temperatures is given a priority, where some values are more important than others. The higher the priority, the more interest the model has in keeping the temperature at a desired setpoint. Lower priority temperatures can be slightly off setpoint without affecting the glass quality.

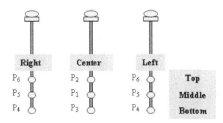

Figure 7: Nine equalizer triplex temperature values

In this new system rather than controlling temperatures by individual zones, the mass flow temperature is used to control the entire forehearth. Individual zone temperatures are no longer of interest, because the equalizer triplex temperatures are now controlled directly. The setpoint for the mass flow temperature is automatically adjusted by the gob temperature model.

FEED FORWARD REAR ZONE DISTURBANCE REJECTION

The rear zone (and sometimes distributor) is used in this strategy to remove any incoming temperature disturbances and to position the glass temperature in a range that allows the front and equalizing zones to handle the final conditioning. The entrance forehearth temperatures are fed forward and used as an input to the forehearth model. The system is now aware of an upset before it reaches the forehearth entrance, and output adjustments can be made to eliminate the temperature upset before it enters the remaining forehearth zones. Ultimately, this prevents unstable glass from traveling through the forehearth.

Figure 8: Feed forward

GOB TEMPERATURE CONTROL

The gob temperature measurement from the GTS is used as the process variable for a model based control loop where an operator can enter a desired gob temperature setpoint. The output of this model determines a new setpoint for the mass flow temperature. The mass flow temperature is a composite of the nine triplex readings in the conditioning zone of the forehearth. The desired MFT is adjusted by the gob temperature control model. Model based controllers are used to adjust zone heating and cooling values to maintain the desired MFT therefore controlling the forehearth as one unit. To avoid process upsets and disturbances before they reach the conditioning zone, there is rear zone disturbance rejection (feed forward). Feed forward models are used to minimize incoming temperature disturbances. These models eliminate the effect of job changes on adjacent forehearth and melter upsets.

RESULTS OF MODEL BASED CONTROL OF THE GOB TEMPERATURE

The following presents the results of the gob temperature model based control system installed on a dual-gob flint glass container line.

Table I. Expected benefits and results of gob temperature control system

Expected Benefits	Results of Field Trial
Reduced variability of gob temperature	7x reduction in gob temperature variation (95% confidence)
Faster compliance to setpoint changes	
Faster recovery from disturbances	Reduced gob temperature stabilization time after a job change from 5-7hrs to 0.5-4 hrs
Improved process quality	
Improved performance after a job change	Significant reduction in lost production volume caused by job changes
Improved gob weight consistency	50% less time required to achieve stability

Gob Temperature Stability

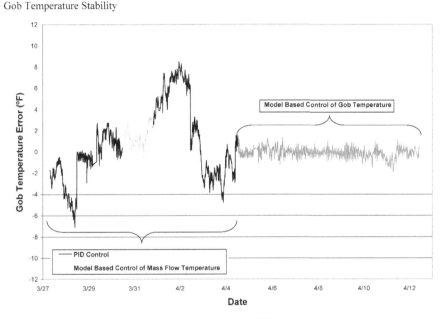

Figure 9: Gob temperature variation for different control methods

Figure 9 displays the gob temperature error on one container line during a 16-day period with the following configurations:

1) PID control
2) Model based control with the mass flow temperature as the process variable
3) Model based control with the gob temperature as the process variable

The gob temperature error is defined as the difference between the desired and actual gob temperature. In the case of the MBC system using the gob temperature measurement as the process variable, the error is the difference between the measured and the setpoint.

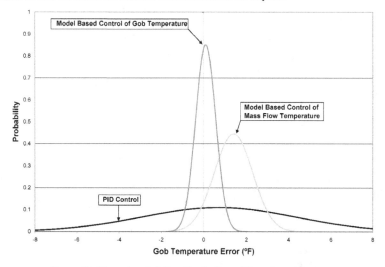

Figure 10: Normal probability distribution of the gob temperature error under different control methods

Figure 10 displays the normal probability distribution of the three control methods during the 16-day period. It can be seen how the model based control system controlling the mass flow temperature achieves greater stabilization of the gob temperature as compared to PID. When the gob temperature is used as the process variable, the model based control system is able to further stabilize the gob temperature and keep it much closer to the desired value.

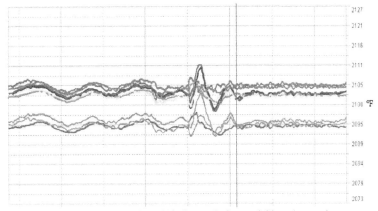

Figure 11: Nine point grid before and after model based control

Job Change Performance

The two figures below show the 9-point grid temperatures during job changes. These two forehearths are essentially equal to one another in size, tonnage, and pull rate, and the same setpoint changes have been made. The results are dramatic and show the ability of model based control to quickly stabilize the temperature during job change, thereby minimizing the time it takes to return to steady operation.

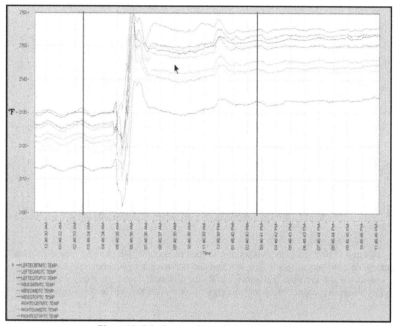

Figure 12: Job change with optimized PID tuning

Figure 12 represents optimized PID tuning, which was tuned as well as possible prior to the change. The total time from setpoint change to stabilization is approximately 8 hours.

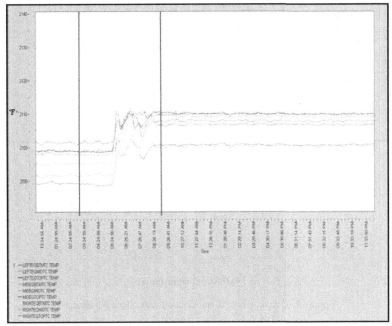

Figure 13: Job change with model based control

Figure 13 represents a job change on a forehearth using model based control. The total time from setpoint change to stabilization is approximately 3 hours, only 25% of the time it took for PID to stabilize.

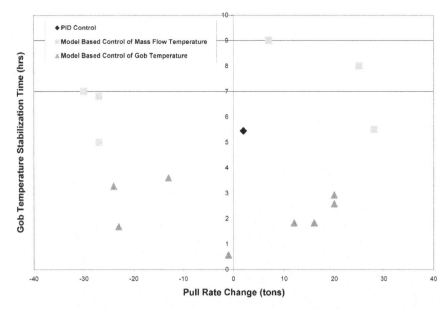

Figure 14: Gob temperature stabilization time after a job change

Figure 14 contains data from fifteen different job changes given different control methods. Both the pull rate change and time required to achieve stable gob temperature conditions are plotted. Stable conditions were defined as ±1°F for the MBC. PID control could not achieve ±1°F so the time required to reach typical gob temperature stability was recorded.

In order to equalize the data displayed in Figure 14, the ratios of tons changed to time required for gob temperature stabilization were calculated. Figure 15 displays the averages of the calculated ratios defined as the gob temperature recovery rate for the different control methods.

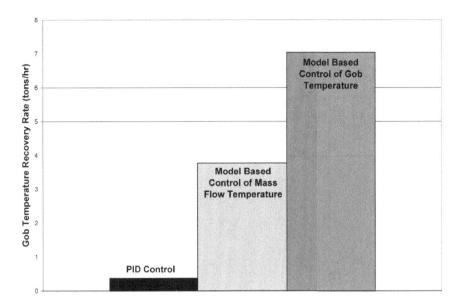

Figure 15: Gob temperature recovery rate after a job change

CONCLUSION

The BASF Exactus advanced gob temperature sensor is capable of providing an accurate and repeatable temperature measurement of each sheared gob. The sensor has demonstrated that with conventional PID control, the gob temperature can vary significantly even though the equalizing thermocouples indicate stable behavior. At the Saint-Gobain Containers, Inc. – Dunkirk, Indiana, USA plant, it was found that if the gob temperature could be maintained to within ±2.5°F of the optimum temperature, the presence of seal surface and split finish defects could be significantly reduced.

The ACSI model based control system (BrainWave[g]) possesses the ability to tightly control the gob temperature to better than ±1°F. The BrainWave system treats the forehearth as a single unit rather than several zones and continuously adjusts the upstream process in order to maintain the setpoint. The BrainWave system quickly reacts to process disturbances and significantly reduces the time required for the gob temperature to return to a stable condition after a job change.

By combining the advanced technologies of BrainWave and Exactus, ACSI and BASF are able to offer container manufacturers a solution that tightly controls the glass temperature at the point of entry into the forming process. This new type of control allows the plant to minimize container defects, improve production efficiency, and significantly reduce the gob temperature recovery time after a job change.

APPLICATIONS AND CHALLENGES FOR INFRARED TEMPERATURE MEASUREMENT IN GLASS MANUFACTURING WITH AN EMPHASIS ON TEMPERING OF LOW-EMISSIVITY GLASS

James Earle and Frank Schneider
Raytek
Santa Cruz, CA

HISTORY AND BASICS OF INFRARED TEMPERATURE MEASUREMENT

Sir William Herschel accidentally discovered the infrared region of the electromagnetic spectrum in the 1830's. Herschel was conducting a series of experiments to determine which color was hottest. Using mercury thermometers and a prism to refract the light into its' constituent colors, he measured the temperature of the different colors. However, during one of these experiments he left a thermometer on the table outside the visual colors and noticed that this thermometer became much hotter. From this he hypothesized that there exists a region of the electromagnetic spectrum just beyond the visible spectrum, which transmits more heat than visible light. This is the infrared region and it is now defined as radiation with wavelength from 750nm to 1mm. Gustav Kirchhoff proposed the law of thermal radiation, which stated that infrared emittance was a unique function of power; essentially a body emitted a unique spectrum and quantity of power as a function of that body's temperature. Finally, Nobel Laureate and father of Quantum Theory, Maxwell Planck, quantified the emission of radiation. Thus, the temperature of any body (above absolute 0) can be determined by measuring the amount of emitted radiation.

Unique Relationship of Temperature and Energy

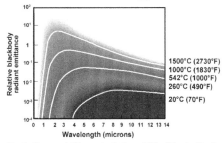

Fig 1. Spectral Characteristics of Blackbody Radiation

In more modern times, instruments have been developed to make use of this phenomenon to measure temperature. In these infrared thermometers, the infrared radiation is collected by the optical system and a lens simply focuses the radiation onto an element, which is sensitive to the radiation. Several types of elements are used, the most common are: thermopile (thermo voltaic), Photovoltaic (photo diodes) and pyroelectric. In this sense, an infrared sensor is not so different from a traditional camera where the lens focuses the image on a light sensitive medium, for a camera the medium is film for the sensor it is an infrared detector. The key parameter here is that in the same way that objects outside the field of view of a camera lens do not appear in the picture, temperatures outside the field of view of the thermometer optics will not be measured.

Another important feature, the infrared sensor measures the complete area within the field of view defined by its optics. Thus, the measurement is an average temperature over the field of view and for accurate measurement the target must completely fill the sensors field of view.

Object Should Fill Field of View

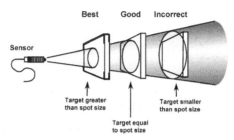

Fig. 2 Targeting Issues

An infrared temperature sensor measures the amount of emitted radiation from a specific wavelength bandwidth, commonly referred to as the spectral response. However, the sensor cannot discriminate in regard to the source of that infrared radiation. Possible sources of radiation from a target are emitted, transmitted and reflected. Since only the emitted power is proportional to temperature, transmitted and reflected radiation constitute potential sources of measurement error. At thermal equilibrium, emitted + transmitted + reflected radiation = unity (E + T + R = 1). The rate at which a body emits energy is referred to as it's emissivity and the rate at which it transmits is referred to as its transmissivity. For opaque targets transmissivity equals zero, thus measured radiation is due only to emitted and reflected radiation. A body's emissivity is often dependent on its temperature and the wavelength of measurement. Thus determination of a body's emissivity must take into account the wavelength at which the measurement will be taken and temperature of the body. The topic of emissivity is often quite confusing, more so the case when the target body is partially transparent. This subject may become a little clearer by studying the graphic in Figure 3 below.

Reflectivity, Transmissivity, Emissivity

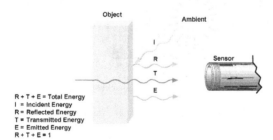

Fig. 3 Sources of Infrared Energy

INFRARED TEMPERATURE MEASUREMENT AND GLASS

Glass is a partially transparent target. Many people will look at glass and say, "glass looks completely transparent to me." However, glass is not transparent at all wavelengths. Our eyes perceive radiation in the wavelengths 400 to 700nm. In this range glass is almost completely transparent. However, at wavelengths beyond ~5 um glass is almost completely opaque. From1 – 5 um the amount of transmission varies depending on glass type. This partial transmittance allows infrared sensors to measure temperatures either on the surface or at some depth into the glass. The depth that the sensor reads into the glass is dependent on both the glass type and the spectral response of the infrared thermometer. In cases, when depth measurements are desired a short wavelength sensor is used. For example, a clear soda glass with thickness of .25" might emit only 10%. So a simplistic approach would determine that if the same soda glass were greater than 2.5" thick, then 100% of the energy would come from the glass and the instrument could measure up to 2.5" in to the glass. The simplistic approach is not too far wrong, and in reality the true measurement of partially transparent glass ends up as an integral equation of energy contributions up to a terminal depth within the glass. This terminal depth is sometimes referred to as the effective depth. In the case of clear soda glass measured with a 1um instrument, the effective depth is typically ~25mm.

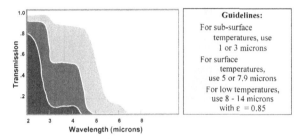

Fig. 4 IR Measurement of Glass

One difficulty associated with the temperature measurement of glass is the fact that glass is an excellent thermal insulator. As a result of this property, thermal stratification happens quite profoundly and rapidly in glass cooling and heating processes. As a result one needs to consider the type of glass being measured, it's thickness and the process to achieve good results. In terms of the type of glass, the key requirement is to know approximately the transmission of the glass, as this will affect effective depth. For extremely thin clear glass, longer wavelengths will be required to make any measurement. Meanwhile, an amber glass may not have the transmission required to gain an effective depth of measurement for "bulk" glass temperatures. Small glass targets being reheated by flames will display much different temperature when measured with a 1um, 2.2um, 3.9um or 5um instrument due to the different effective depths and thermal stratification at those different depths. In these types of applications, it is usually desired to measure as close as possible to the heating processes in order to better track surface heating. This is not case with bulk temperature applications. In these applications, a significant amount of glass is present and the surface temperature can be either hotter or cooler due to surface exposure with atmosphere. In these situations measurement at depth is desired to overcome temperature gradients and produce a "bulk" temperature.

LOW EMISSIVITY GLASS

Low-emissivity (low-e) glass refers to glass that has a special coating applied to it so as to control heat transfer through the window. Typical coatings are microscopically thin layers of metal or metal oxides, almost invisible to the naked eye. These have the effect of allowing most of the visible light to pass through the glass but will block wavelengths in the infrared range that are more inclined to transfer heat energy.

In commercial and domestic glazing applications, the use of low emissivity glass has been proven to significantly improve the energy efficiency of the structure. While low emissivity glass has existed for many years, heightened consumer environmental concern has driven a rapid growth in this type of glass. Many municipalities now enforce the use of low-e glass through building codes, further fueling the growth.

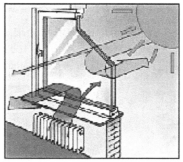

Fig. 5 Low Emissivity Glass in Architectural Applications

Types of low -e coatings are typically referred to as "hard-coats" or "soft-coats". Hard-coats, sometimes referred to as pyrolytic coatings, are typically formed by a Magnetron Sputtered Vapor Deposition (MSVD) process and are have excellent durability and moisture resistance properties. Soft-coats are typically applied by a Chemical Vapor Deposition (CVD) process and are characterized as being less durable and having less moisture resistance. Both are available in many grades from many different suppliers leading to a wide variety of properties.

TEMPERED GLASS

In the tempering process, glass is heated above its annealing point (typically about 600°C) and then rapidly cooled in an air quench process. This results in greatly increased internal stresses and the glass physical properties are altered. The glass is considerably harder after this process and will have a tendency to break into multiple small pieces rather than large shards. Tempered glass is used in areas where such large shards could cause injury in the event that the glass panes were to break. Application examples include glass doors and windows below a certain height. As is recently the case with low-e glass, the use of tempered glass in building is frequently mandated by building codes.

As one can imagine, since exposing it to high temperature and then rapidly returning it to a lower temperature alter the glass properties, measurement and control of temperature is a key parameter of the glass tempering process. Any typical contact temperature measurement methodologies such as thermocouples or RTD probes are impractical and IR temperature measurement has been the default methodology for several years. To achieve the highest quality, the desire is to heat the glass sheet so that it not only is heated to the right temperature but also has an even heat distribution across the sheet. To accomplish this, typically an IR linescanner is mounted above the process looking down at the glass sheet as it exits the oven. Incorporating an IR detector in the 4.5 to 5.2μm spectral response range, the linescanner will return 256 data points across the glass sheet surface each time it scans the target. Typical scanning speeds would be in the 50 to 100 Hz range. Since the glass is moving, each line scanned is incrementally separated from the last line. By combining these images on a PC, a two dimensional false color thermal image is created. Application specific PC software has been developed to allow the user to define zones across the sheet, alarm conditions and to conduct comprehensive data analysis trough a range of databasing and archival tools. Systems of this type have become a key tool in the challenge to provide higher throughput and quality in the glass tempering process.

Fig. 6. IR Linescanner in Glass Tempering Application

CHALLENGES OF TEMPERING LOW-EMISSIVITY GLASS

While the process to measure temperature in the glass tempering process has been well proven for standard flat glass, low-e glass presents some unique challenges.

Many of the ovens used in the tempering process utilize IR energy as a means to heat the glass. Since the low-e coating is specifically designed to reflect this radiation, low-e glass will not heat up at the same rate as non-coated glass. This means that process settings need to be established and maintained for each type of low-e glass processed.

Another concern is that the standard IR linescanner system requires the scanner to be mounted above the process looking downward. The width of scan required dictates that the scanner be a considerable distance away from the glass and, short of preparing a hole in the factory floor, the only viable mounting place is above the process. However, since abrasion caused by the conveyor rollers can damage the low-e coatings, this type of glass is always tempered with the coated side facing upwards. Thus the glass surface seen by the IR linescanner is the surface that has been specially altered to have low emissivity and thus be a less than ideal candidate for IR temperature measurement. Without correction, the linescanner will read temperatures to be considerable lower than the actual process temperature.

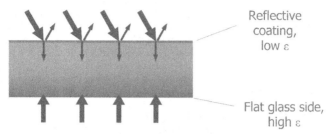

Fig 7. Orientation of Glass in Tempering Oven

THE SOLUTION

A unique and novel concept has been developed which utilizes temperature measurement on both sides of the glass sheet. A single-point sensor is located underneath the process and the IR linescanner is mounted above. Both data streams are routed back to a PC via RS485 communication and processed by the application specific software. During an initial configuration process, the physical locations of both sensors are entered. The system can then identify which of the 256 data points from the linescanner is from the same coordinate as the single point sensor. Once this information is provided, the software compares the two measurements, one from the point sensor and the other from the linescanner, which are taken on opposite surfaces of the same point on the glass sheet. The underside of the glass is uncoated and thereby the data gathered is the "true" temperature while the upper side is highly reflective in the IR range and the data is not useful without emissivity correction. The two values are compared and the emissivity is calculated. This emissivity value is then applied to all 256 data points and a corrected thermal image is displayed. By applying this correction factor the system essentially becomes independent of changing emissivity and therefore, it can automatically adjust itself when glass type changes.

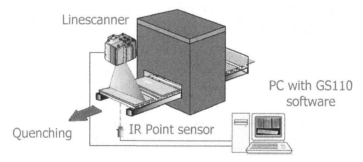

Fig 8. Unique System for Temperature Measurement when Tempering Low-E Glass

SYSTEM FEATURES

A number of specific challenges had to be overcome in the development of this system. The first is to accommodate a situation where a smaller glass sheet may pass under the scanner but may be small enough to pass to one side of the point sensor underneath the production line. In this event there is no reading from the point sensor and therefore the correction algorithm is no longer valid. To counteract this situation, when there is no reading from the point sensor the software is designed to use the last valid emissivity value. Therefore, as long as this glass sheet is of the same type as the sheets before it, a valid emissivity setting is used and a corrected image is displayed.

Another challenge is in the area of system cooling. Since the glass sheets have to be heated in the region of 650°C the ambient temperature in the area where the glass leaves the over is obviously quite high. Since the linescanner is mounted directly above the process, it is clearly in a very vulnerable position. This situation is made worse in the event of a glass break inside the oven, where the oven door must be opened wide to clear the blockage; exposing the scanner to even more heat. The scanner itself is equipped with internal cooling coils that will allow it to operate in ambient temperature up to 150°C when connected to cooling water. For locations where water is not acceptable, a special enclosure was designed with an integral heat shield. The refractory lined shield limits the heat reaching the enclosure and a high performance vortex cooler actively cools the enclosure. The point sensor beneath the glass is not immune from exposure to heat. In this case, an air collar keeps the sensor cool while a cooling hose is used to protect the cable.

SYSTEM LIMITATIONS
The algorithm used to correct the emissivity uses a methodology that takes the temperature values from the top and bottom of the glass sheet and calculates the supposed emissivity. This emissivity is used on the next sheet and the error is examined. If too large, another correction is applied and so on until an accurate result is returned. This is in principal an iterative process and there fore it requires that several glass sheets be processed before an accurate value is returned. Within the system configuration, the operator has an opportunity to enter the expected emissivity of the target. If the value entered is close to the actual value then the correct emissivity can be reached after just one pass. If there is a significant deviance or if no value is entered, it will take several passes before the image is properly corrected. Typical performance is three passes, even if no value is entered. This is more than acceptable in most situations, as most tempering facilities operate in batch mode, i.e. one type of glass at a time. If the user wanted to process mixed glass types, the system could not correct the images in real time.

What we know is that the system will provide high quality images, corrected for emissivity in all cases where the glass in question has an emissivity of 0.3 or higher. As the glass emissivity reaches lower values, the laws of physics present significant problems. The energy reaching the scanner from the coated glass becomes extremely low. In cases where the emissivity is between approximately 0.1 and 0.3, the images will be displayed but "noise" becomes evident and image quality suffers. Also at these lower levels, the data can be affected by reflections of other heat sources such as lamps etc., in the glass surface. Shielding the scanner and the target will mitigate this and allow the system to present better results at lower emissivity values. When the glass has an emissivity below 0.1, there is almost no energy reaching the scanner. The image will show extreme noise and the emissivity may not correct consistently. When the emissivity reaches 0.05, the system will no longer correct the images.

SUMMARY
Infrared thermometry has as its genesis the early experiments in fundamental physics. Since then it has been developed and refined to the point where today, IR thermometers are found in a hugely diverse sampling of industrial processes. Continued refinements have converted this interesting phenomenon into a "must have" tool for any process involving heat.

The increased use of low-emissivity glass in the building industry has brought with it an increase in the challenges associated with its tempering. The system described in this paper is the first known attempt to address these problems and the only known commercially available system of its type. While the system has some `limitations in that in cannot function on glass where the emissivity is approaching zero, it has proven to be a useful tool to increase throughput, improve throughput and reduce cost.

INDUSTRIAL EXPERIENCES WITH A NEW SURFACE TREATMENT TECHNOLOGY

Heiko Hessenkemper
TU-Bergakademie
Freiberg, Germany

ABSTRACT

A short time contact of Al with a hot glass surface above the transition temperature results in surface changes of the glass composition. In addition structural changes with complex interactions take place. An improvement of the chemical resistance is reached and leads into another hydrolytic class. Mechanical properties like scratch resistance, bending strength and impact strength are optimized. The transfer into the industrial production situation is realised using $AlCl_3$ vapour or dissolved $AlCl_3$ in alcohol which is burned in a reaction room in an online process after the forming of the glass. The use of Aluminium moulds as a third route is still under development.

INTRODUCTION

Glass industry is suffering from a strong competition and rising energy prices increase the problem. The classical glass industry chooses the glass composition as a compromise between different demands: Basically the customer demand has to be fulfilled, but economic and technical demands resulting from the batch and furnace area as from the forming side have some influence towards the glass composition, too. One possibility to optimize the process is to separate the different resulting profiles and demands, because a lot of properties are defined by the glass surface. Therefore by changing the surface properties there are two basic possibilities:

1.) To improve the properties and open up new markets
2.) To gain the same properties with decreased costs, for instance using a different glass composition

This approach is not new. Surface property changes like thermal hardening and coatings are in industrial use since decades with great success.

We have searched for a cheap new process which should be suitable for the mass glass production and have found the use of $AlCl_3$ as an interesting option leading to a technique being sheltered by an international patent [1].

Laboratory results have shown that it is possible to increase the chemical resistance more than 10 times. The mechanical properties like bending strength could be improved for more than 50 %, the scratch resistance was improved, the reflection was decreased with an increased transmission and the surface electrical conductivity was decreased. Most of these effects could be explained by an incorporation of Al into the glass matrix which was shown in diagram 1, where a strong increase of Al up to a depth of 50 nm is evident.

Diagram 2 demonstrates the strong improvement of the hydrolytic resistance of container glass with an improvement of approximately 900 %. An interesting aspect is the observation, that the Al release is strongly reduced even with a great Al enrichment in the surface which opens up interesting discussion about the possible processes which happens with this surface changes.

Another important aspect coming from laboratory tests has been the fact that this surface improvement technique works for all kind off glass compositions: Soda lime glass, lead crystal glass, borosilicate glass and even with different enamel compositions.

With $AlCl_3$ as a cheap material and only small amounts being used for the enrichment of the glass surfaces the question arises, how this method could be transformed into an industrial technique.

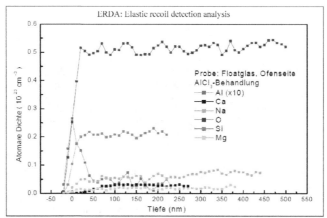

Diagram 1: Element analysis of treated float glass

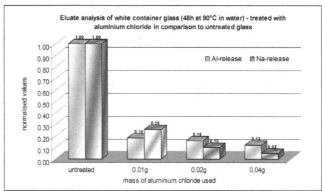

Diagram 2: Change of hydrolytic resistance of treated container glass

INDUSTRIAL EXPERIENCES:
So far industrial experiences have been made in the area of flat glass, tableware, container glass, tube glass and fibre glass. The process works with some differences for soda-lime glass, borosilicate glass, lead crystal glass and different glass compositions being seen in the field of enamels and glazes.

The transfer into the production process has always to optimise the following basic aspects controlling the process:

1.) Temperature(Viscosity) of the glass surface
2.) Reaction time
3.) Concentration of the AlCl$_3$

Different technological solutions for different glass productions have to be found and some results are discussed.

A first basic result already being seen from laboratory results is that the improvement concerning the chemical resistance demands higher concentration, temperatures or reaction times than the optimised values of the mechanical improvement are asking for. The interpretation of this result is complicated and has to considerate several aspects being influenced by the surface changes of the glass.

Concerning the appropriate technology it turned out that the use of direct AlCl$_3$ vapour is more complicated although a sublimation temperature of about 180 °C seems to make it easy to create the vapour. But the storage, the transport to the reaction chamber and other difficulties makes it more comfortable to dissolute AlCl3 in alcohol and burn it in a reaction chamber, creating the necessary atmosphere situation direct on the hot glass surface. In any case the exhaust fumes have to be separates and treated due to aggressive materials as HCl and AlCl$_3$ which has not been reacted. The humidity and oxygen partial pressure are additional parameters which could influence the results.

FIBRE GLASS:
For fibre glass a special problem came up: An extreme short reaction time with the creating of new glass surfaces at the same time. Therefore half industrial equipment with 200 nozzles at the Institute of Polymer Research in Dresden has been used. The results are shown in diagram 3:

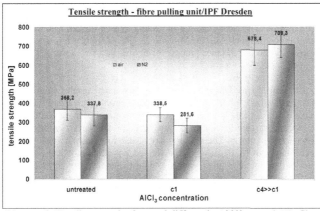

Diagram 3: Tensile strength of several differently AlCl3 treated AR- fibre glasses without sizing (measurement IPF)

The uncoated fibres reach about 100% higher tensile strength proving the principal possibilities which has been realised in laboratory results. The problem was the reproducibility of the process which has not been solved with this equipment. Therefore we are going back to a mono fibre equipment to study the process more in detail.

TUBE GLASS:
      Laboratory results with a batch treatment of industrial glasses are shown in diagram 4:

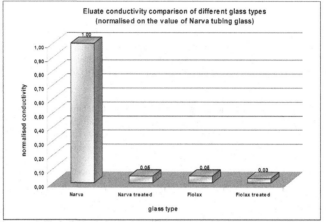

Diagram 4: Hydrolytic resistance of different treated and untreated tube glasses

The Narva tube glass is a lamp glass with a low chemical resistance. It could be improved to the same chemical resistance as Fiolax, a brand name for a borosilicate glass being used for instance for pharmaceutical applications. It is obvious that the Narva glass with perhaps 30 % of the production cost reaches similar properties and even the borosilicate glass could be improved. We transferred the inside treatment into a Vello and Danner process, using instead of air for the internal pressure AlCl$_3$ vapour.

For both, Vello (diagram 5) as Danner process the laboratory results concerning the hydrolytic resistance could be repeated in the industrial process, but a strong decrease of the mechanical strength has been seen. The reason for this is a change in the surface composition combined with a different extension coefficient and surface tension leading to a rough and surface with decreased mechanical properties being seen in diagram 6. The effect is strong due to the low viscosity of the glass surface coming in contact which AlCl$_3$. Because of the limited possibilities to vary the parameters in the process the future will be for tube glass not the online treatment, but the batch treatment after the production, for instance after the forming of pharmaceutical packaging. The economic perspectives with changing the glass system but keeping the same quality standards or reaching new quality standards are huge and could be a strategic tool in the market.

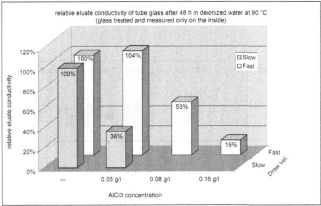

Diagram 5: Vello process

Diagram 6: Rough surface after $AlCl_3$ treatment in an industrial Vello process (pictures taken by SEM)

Another aspect has to be discussed: For certain glasses special Al containing raw materials are used to improve the chemical resistance and the mechanical properties. But these are surface defined properties. By optimising the surface properties with $AlCl_3$ there is no need for this special raw material. This could, depending on the special situation of a plant, save up to 3 €/ton of glass in raw materials. In addition the refining temperature could be reduced with energy savings, reduction in CO_2 emissions and lifetime increase of the furnace. These aspects are valid for all kinds of glass production.

CONTAINER GLASS

In container glass several industrial tests has been conducted. Diagram 7 shows a result using bulk material of $AlCl_3$.

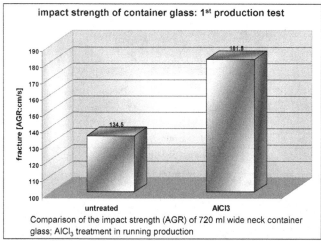

impact strength of container glass: 1ˢᵗ production test

Comparison of the impact strength (AGR) of 720 ml wide neck container glass; $AlCl_3$ treatment in running production

Diagram 7: Bulk $AlCl_3$ in container glass production

For a strong treatment with a high chemical resistance it could be necessary to wash the glass due to a fog of NaCl which is created on the surface. With a right concentration it is possible to substitute the hot end treatment. Therefore a hood has been constructed as a reaction chamber to have a better control for the parameters. The tests are still going on.

FLOAT GLASS

For float glass the online treatment is under development. But in addition the batch treatment is a possibility with a great potential. Diagram 8 shows the results for an autoclave test concerning the optical properties which has an important impact for the strong growing photovoltaic industry. But in addition the improved mechanical properties open up new possibilities in the thermal hardening of float glass. A project has been started to reach 2 mm and less thermal hardened glass for automobile and architectural glasses where at the moment the border line is at about 2.8 mm thermal hardened glass. This will in return open up new strategic markets.

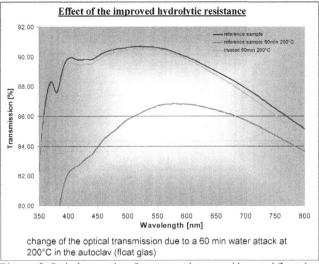

Diagram 8: Optical properties after an auto claw test with treated float glass

TABLEWARE

For tableware and flaconnage several industrial tests have been performed developing a reaction chamber being integrated on the transportation belt just after the forming process. Diagram 9 shows the results for the impact strength and the hydrolytic resistance for perfume bottles treated online under different parameters. Nr. 0 and 10 are the untreated material.

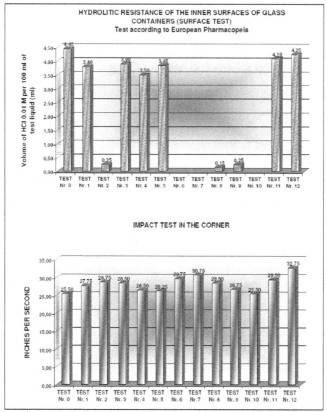

Diagram 9: Perfume bottles with different online treatment of AlCl₃.

For stemware similar results with an online treatment after the forming process have been achieved similar to the laboratory results (Diagram 10). Number 0 is the untreated material.

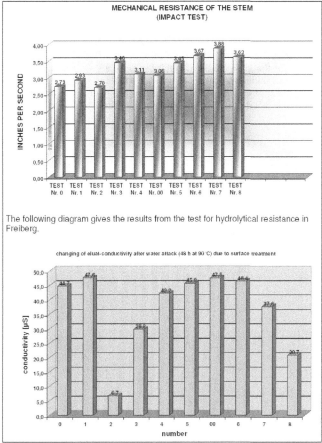

Diagram 10: Stemware with different online treatment of AlCl₃.

From this these tests a not only the practical experiences have been achieved but in addition rough cost estimation for tableware has been possible: Beside the cost for the license fee an investment in the range of 10.000 € per production line is necessary and running costs of about 20 €/ day after further improvement of the process. Compared to these costs the above mentioned possible savings, the improved quality and new market chances have to be accounted.

CONCLUSION

The new surface treatment technology using AlCl₃ in contact with hot glass surfaces has demonstrated the practical use and it has been possible to transfer the laboratory results to several industrial tests. A cost-result analysis has proofed the strategic impact of this technology: Either to improve with existing glass compositions the product properties or to have similar properties with changed glass compositions resulting in interesting cost reductions. Both aspects can open up new markets. The process to transfer the technology is still going on, facing in special glass productions certain problems which have to be solved in the near future.

This new perspective for the glass industry is probably again a good example, that the problems of glass industry will find a solution basically in technological tools and a sticking to just controlling tools will only extend the dying of an industry. On a long term the market will decide about the different philosophies and we hope that it is not too late to change the attitude towards new technologies.

Literature:
[1] Patent Nr. PCT/EP2004/004642, WO 2004/096724 A1: Alkaline Glasses with Modified Surfaces and Method for Producing Same

WHAT IS THE IDEAL GLASS BATCH DEPTH OF A GLASS FURNACE?

E. Muijsenberg
Glass Service B.V.
Maastricht, Netherlands

M. Muijsenberg and J. Chmelar
Glass Service Inc.
Vsetin, Czech Republic

Glenn Neff
Glass Service USA Inc.
Stuart, FL, USA

ABSTRACT

Mathematical modelling is reaching a high acceptance level within the glass industry. Today most new furnaces are being modelled before the final design is decided. It is clear that the modelling helps to optimise the furnace in respect to glass quality, energy efficiency and furnace life-time. The extra effort of the modelling is leading for sure to a quick pay-back of this extra investment and an increased profit over the furnace life-time. Even the furnace life-time can be extended with better insight on temperature distribution and glass speeds that corrode the refractory. Many glass produces are always asking us: "what is the optimal glass depth"? There is not just one answer to this, but the paper demonstrates how mathematical modelling can help to find the optimal furnace depth for a certain furnace design.

1. INTRODUCTION

With mathematical modelling we divide a furnace in small volume elements and calculate over these volumes the conservation equations for mass, energy and momentum. The result is that we obtain a flow and temperature distribution of the glass (and combustion gases) within the glass furnace. As a second step with post processing tools we can predict these days what will be relative trends with regard to the glass quality and production yield, next to furnace emissions and energy efficiency. We can estimate that more than 60% of furnaces being designed and constructed today have been optimised before with some mathematical modelling tool. Only Glass Service already executes about 40 of these optimisation studies per year in-house. Our customers using our license are estimated to make in total about 120-150 furnace design optimisation studies per year.

2. THE BASE CASE

For the purpose of this study we used a non existing but very typical container glass furnace design. The glass is clear soda lime glass using 30% cullet. The furnace in this example is all the time melting 180 Tons per day on a surface of 74 m2, so 2.43 tpd/m2. The base case design glass depth is 1.4 meter. The total fuel input was 900 Nm3/hr. So specific energy use was about 4 MJ per kg of glass. The combustion is fired by 2 underport gas injectors injecting the natural gas into the regenerative preheated airstream.

The following figure 1 will show us a 3D view from the top looking to glass surface and the flame generated above. The colors shows temperatures, from blue around 1000 °C till red 1800 °C.

What is the Ideal Glass Batch Depth of a Glass Furnace?

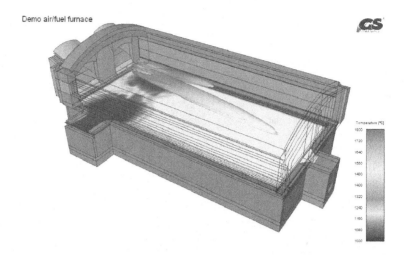

Figure 1. 3D view of air fired regenerative furnace.

The next 2 figures show a vertical cut through the length of the furnace. Figure 2, shows a plane right at the center, showing the also the throat. The next figure 3 is shifted to the side showing the flame development on the firing side of this U-flame furnace.

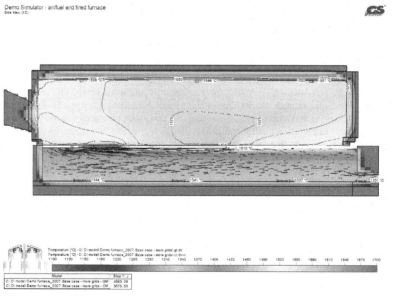

Figure 2. Side view in center plane of furnace.

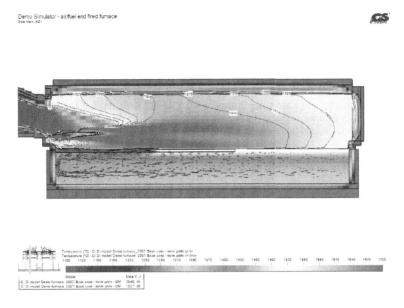

Figure 3. Side view to the side near firing side of furnace.

3. THE CASES

From this base case situation we calculated some variations to answer one of the most popular questions we hear repeatedly when we optimise the design of a new furnace:

"What is the optimal depth for this glass furnace?"

Now we have to say that of course the optimal depth depends on several factors. Most important here is the glass transmission, e.g. - clear, green or amber. But there are several other variations here, an oxidized green glass does not have the same effective thermal conductivity as a reduced green glass will have.

Beside the glass type of course the ideal glass depth is also a variation of tonnage, firing principle, bubbling (or no bubbling), electric boosting, barriers etc etc.

So in this paper we just demonstrate for a basic U-flame furnace, melting clear glass what is the optimal depth for this design and tonnage.

That means the conclusions are only valid for this furnace and this glass type and fuel rate etc. The conclusions might change completely when for instance doing the same study for green glass instead of clear glass.

We tested how the furnace would work when the glass depth is decreased or increased with steps of 20 centimeters. Note that for darker glass the effect will be stronger than for clear glass (the darker glass will have a higher temperature gradient)

Figure 4 shows the results of the glass bath temperature for all five cases, starting from the most shallow on top and the deepest one on the bottom. We can clearly see that the more shallow furnace has higher bottom temperatures and less recirculation than the deeper ones. The colours represent the temperature. Dark blue is 1300 C and red is 1500 C.

We can see how the shallow furnace has the highest bottom temperature, on one side this is good as it is good for melting conditions, however the bottom **refractory corrosion** will be enhanced too. This will have a negative effect on lifetime of the furnace but also on glass quality (eg more stones). If we compare the temperatures also with the velocity profiles in figure 6 we can see how the recirculation is reduced in the shallow furnace. On the other hand in the deeper furnaces the last 20 till 40 centimeters actually are not participating much in the melting process. They just contain certain amount of "dead glass" that might spoil the good glass when having unstable conditions. Especially the doghouse corners in the most deep furnace are getting very quiet and so relatively very cold.

When you look at figure 5 you can see that the most shallow case starts to behave more as a heating pipe, the temperatures over the bottom increase. Also there is not enough free glass recirculation to heat the area underneath the batch. Also the deepest furnace starts to be more colder underneath the batch.

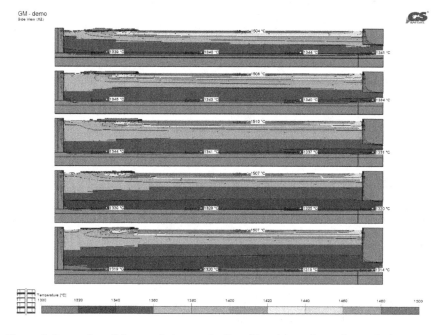

Figure 4. Comparing all 5 cases below each other. The picture shows the vertical temperature distribution in the center of the furnace.

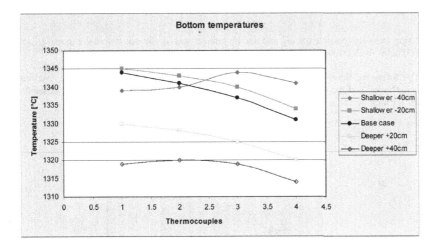

Figure 5. Shows the bottom temperatures of all cases.

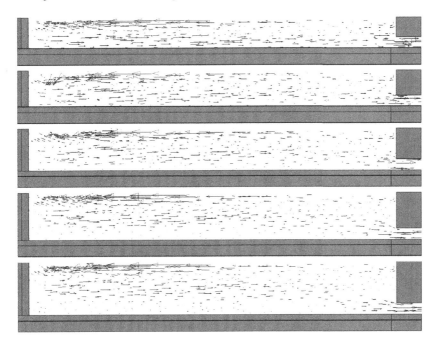

Figure 6. Velocity profile comparison of all 5 cases, on top the most shallow one and on bottom the deepest furnace

4. RESULTS

Using mass less particle tracing and tracing bubbles who really can fine in the model (reach glass level) helps us to evaluate what are the good and what are the less good cases.

The following figures show us the relative changes of minimum residence time, melting index, fining index and the mixing index. For all of them how higher the value is how better the conditions for melting and quality are.

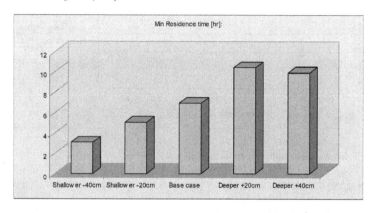

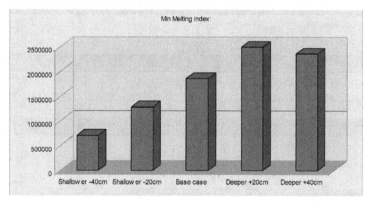

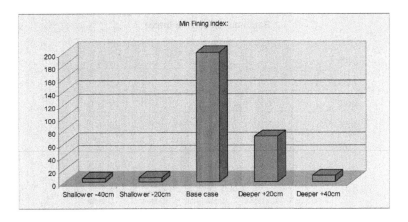

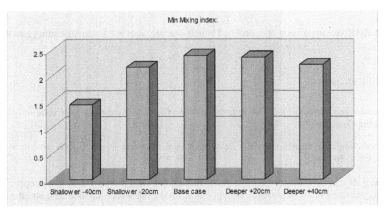

Figure 7. Trends showing all quality indicators

From the quality indicates we can see that the minimum residence time increases with a deeper glass melt, due to the higher volume. The most shallow furnace has only a minimum residence time of 3 hours, which is much too low to have good quality glas (See also the fining index is very low). For the deepest furnace the minimum residence time does not increase further as the lower layers are behaving as dead zones.

The fining index shows us the optimum for the base case deisn and the melting index shows the optimum for the 20 cm deeper furnace.

Last but not least we calculated the redox distribution in the furnace and than traced bubbles. Figure 8 shows us clear trends when tracing actual bubbles that are growing and shrinking during the calculation and can leave the glass melt via the surface or come into the throat. Of course this calculation is more reliable than the relative simple fining index, but in general costs a little more effort to do.

What is the Ideal Glass Batch Depth of a Glass Furnace?

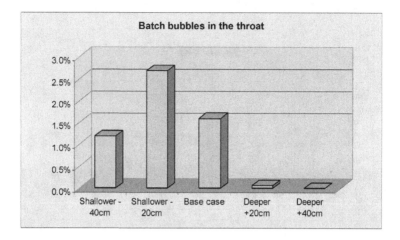

Figure 8. Relative comparision on calculated bubble starting from batch that are ending into the throat

5. CONCLUSIONS

Let us remark again that this is just a demonstration furnace and these results are valid only for this case melting clear glass.

When evaluating the results from the calculation and visaulise in the figures above, we may conclude the following:

Comparing all quality indicators we can conclude that the base case or the 20 cm deeper furnace are actually the best options, probably the optimum is lying inbetween the base case and the deeper + 20 cm furnace.

So for this furnace design, clear glass and this tonnage the best glass quality will be probably be produced with a furnace that is around 1.5 meter deep. At the same moment the lower bottom temperatures will also help the furnace life time and reduce defects potential from the bottom.

One must not forget that such optimisation is furnace design and case dependent and might give different results under different conditions.

There are of course also other options to consider, such as a multi level furnace with steps along the length (or even barriers), this already has been common for some float furnaces (being less deep towards the end) or container furnace with having deeper refiners than the main melting part.

In future we plan to produce some more examples that can give general guide lines to people that design furnaces.

EFFECTS OF SURFACE STRUCTURE ON WETTING OF PATTERNED SUPERHYDROPHOBIC
SURFACES

Bharat Bhushan and Yong Chae Jung
Nanotribology Laboratory for Information Storage and MEMS/NEMS
The Ohio State University
Columbus, OH, U.S.A.

ABSTRACT
    Superhydrophobic surfaces have considerable technological potential for various applications
due to their extreme water-repellent properties. When two hydrophilic bodies are brought in contact,
any liquid present at the interface forms menisci, which increases adhesion/friction and the magnitude
is dependent upon the contact angle. The superhydrophobic surfaces may be generated by the use of
hydrophobic coating, roughness and air pockets between solid and liquid. The geometric effects and
dynamic effects, such as surface waves, can destroy the composite solid-air-liquid interface. Studies on
silicon surfaces patterned with pillars of varying diameter, height and pitch values and deposited with a
hydrophobic coating were performed to demonstrate how the contact angles vary with the pitch. A
criterion was developed to predict the transition from Cassie and Baxter regime to Wenzel regime,
considering water droplet size as a parameter on the patterned surfaces with various distributions of
geometrical parameters. The trends are explained based on the experimental data and the proposed
transition criterion.

INTRODUCTION
    One of the crucial surface properties for various surfaces and interfaces in wet environments is
non-wetting or hydrophobicity. Wetting is characterized by the contact angle, which is the angle
between the solid and liquid surfaces. If the liquid wets the surface (referred to as wetting liquid or
hydrophilic surface), the value of the contact angle is $0 \leq \theta \leq 90°$, whereas if the liquid does not wet
the surface (referred to as non-wetting liquid or hydrophobic surface), the value of the contact angle is
$90° < \theta \leq 180°$. A surface is considered superhydrophobic if θ is greater than $150°$. These surfaces are
water repellant. These surfaces with low contact angle hysteresis (difference between advancing and
receding contact angles) also have a self cleaning effect, called "Lotus Effect". Water droplets roll off
the surface and take contaminants with them.[1-3] They have low drag for fluid flow and low tilt angle.
The self cleaning surfaces are of interest in various applications, including self cleaning windows,
windshields, exterior paints for buildings, navigation ships, utensils, roof tiles, textiles and reduction of
drag in fluid flow, e.g., in micro/nanochannels. When two hydrophilic surfaces come into contact,
condensation of water vapor from the environment forms meniscus bridges at asperity contacts which
lead to an intrinsic attractive force.[4-9] This may lead to high adhesion and stiction. Therefore,
hydrophobic surfaces are desirable. Hydrophobic surfaces can be constructed by using low surface
energy material coatings such as polytetrafluoroethylene or wax, by increasing surface area by
introducing surface roughness and/or the creation of air pockets. Air trapped in the cavities of a rough
surface results in a composite solid-air-liquid interface, as opposed to the homogeneous solid-liquid
interface.[3, 10-14]
    Examples of such surfaces are found in nature, such as Nelumbo nucifera (lotus) and Colocasia
esculenta[15,16], which have high contact angles with water and show strong self-cleaning properties
known as the 'lotus effect,' Fig. 1.[17] Lotus is known to be self cleaning to prevent pathogens from
bounding to the leaf surface. Many pathogenic organisms - spores and conidia of most fungi - require
water for germination and can infect leaves in the presence of water.[15] Recent studies have been carried
out to fully characterize the hydrophobic leaf surfaces at the micro- and nanoscale while separating out

the effects of the micro- and the nanobumps, and the hydrophobic compounds, called waxes on the hydrophobicity.[17,18] The wax is present in crystalline tubules, composed of a mixture of aliphatic compounds, principally nonacosanol and nonacosanediols.[19] By learning from what is found in nature, one can create roughness on various materials and study their surface properties, leading to successful implementation in applications where water repellency, fluid flow and lower meniscus is important.

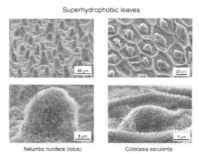

Superhydrophobic leaves

Nelumbo nucifera (lotus)   Colocasia esculenta

Figure 1. SEM Micrographs of two hydrophobic leaf - Nelumbo nucifera (lotus) and Colocasia esculenta.[3]

In this paper, numerical models which provide relationships between roughness and contact angle are first discussed. Also, we present a study of patterned Si surfaces with cylindrical pillars of different diameters, heights and pitch distance in order to investigate the dependence of the contact angles on the geometrical parameters. The experimental data are compared with the transition criterion developed to predict from Cassie and Baxter regime to Wenzel regime on the patterned silicon surfaces with different pitch vales.

CONTACT ANGLE ANALYSIS

Consider a rough solid surface with a typical size of roughness details smaller than the size of the droplet. For a droplet in contact with a rough surface without air pockets, referred to as homogeneous interface, the contact angle is given as[10]

$$cos\,\theta = R_f\,cos\,\theta_0 \tag{1}$$

where θ is the contact angle for rough surface, θ_0 is the contact angle for smooth surface, and R_f is a roughness factor defined as a ratio of the solid-liquid area to its projection on a flat plane. The dependence of the contact angle on the roughness factor is presented in Fig. 2(a) for different values of θ_0, based on Eq (1). The model predicts that a hydrophobic surface ($\theta_0 > 90°$) becomes more hydrophobic with an increase in R_f and a hydrophilic surface ($\theta_0 < 90°$) becomes more hydrophilic with an increase in R_f.[3]

For a rough surface, a wetting liquid will be completely absorbed by the rough surface cavities while a non-wetting liquid may not penetrate into surface cavities, resulting in the formation of air pockets, leading to a composite solid-liquid-air interface as shown in Fig. 2(b). Cassie and Baxter[11] extended Wenzel equation for the composite interface, which was originally developed for the homogeneous solid-liquid interface,

$$cos\,\theta = R_f\,cos\,\theta_0 - f_{LA}(R_f\,cos\,\theta_0 + 1) \tag{2}$$

where f_{LA} is fractional flat geometrical area of the liquid-air interface under the droplet. This model

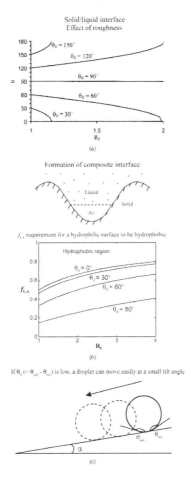

Figure 2. (a) Contact angles for rough surfaces (θ) as a function of the roughness factor (R_f) for various contact angles of the smooth surface (θ_0), (b) formation of a composite solid-liquid-air interface for rough surface and f_{LA} requirement for a hydrophilic surface to be hydrophobic as a function of the roughness factor (R_f) and θ_0, and (c) tilted surface profile (the tilt angle is α) with a liquid droplet).[3]

shows that for a hydrophilic surface, contact angle on a smooth surface increases with an increase of f_{L1}. When roughness factor increases, the contact angle decreases but at a slower rate, due to formation of the composite interface. At a high value of f_{L1}, surface can become hydrophobic; however, the value required may be unachievable or formation of air pockets may become unstable.

For the hydrophobic surface, contact angle increases with an increase in f_{LA} both for smooth and rough surfaces. Using Eq (2), f_{LA} requirement for a hydrophilic surface to be hydrophobic can be found as[3]

$$f_{LA} \geq \frac{R_f \cos\theta_0}{R_f \cos\theta_0 + 1} \qquad \text{for } \theta_0 < 90° \qquad (3)$$

Figure 2(b) shows the value of f_{LA} requirement as a function of R_f for four surfaces with different contact angles, θ_0. Hydrophobic surfaces can be achieved above a certain f_{LA} values as predicted by Eq (3). The upper part of each contact angle line is hydrophobic region. When R_f increases, f_{LA} requirement also increases.

Another important characteristic of a solid-liquid interface is the contact angle hysteresis (θ_H) which is the difference between the contact angle at the increased droplet volume (advancing contact angle, θ_{adv}) and the contact angle at the decreased droplet volume (receding contact angle, θ_{rec}) for a droplet on the solid surface. The contact angle hysteresis occurs due to surface roughness and heterogeneity. Low contact angle hysteresis results in a very low water roll-off angle, which denotes the angle to which a surface may be tilted for roll-off of water drops.[1,2] Low water roll-off angle is important in liquid flow applications such as in micro/nanochannels and surfaces with self cleaning ability.

There is no simple expression for the contact angle hysteresis as a function of roughness; however, certain conclusions about the relation of the contact angle hysteresis to roughness can be made. In the limiting case of very small solid-liquid fractional contact area under the droplet, when the contact angle is large and the contact angle hysteresis is small Eq. (2) is reduced to

$$\theta_{adv} - \theta_{rec} = \sqrt{1 - f_{LA}}\, R_f \frac{\cos\theta_{r0} - \cos\theta_{a0}}{\sqrt{2(R_f \cos\theta_0 + 1)}} \qquad (4)$$

For the homogeneous interface, $f_{LA} = 0$, whereas for composite interface f_{LA} is not a zero number. It is observed from Eq. 4 that for homogeneous interface, increasing roughness (high R_f) leads to increasing the contact angle hysteresis, while for composite interface, an approach to unity of f_{LA} provides with both high contact angle and small contact angle hysteresis.[3,20] Therefore, the composite interface is desirable for superhydrophobicity.

EXPERIMENTAL DETAILS

Single-crystal silicon (Si) was used in the study. Silicon material has traditionally been the most commonly used structural material for micro/nanocomponents.[21] Hydrophilic surfaces can be produced by using silicon material. The Si surface can be made hydrophobic by coating with a self-assembled monolayer (SAM). One of purpose of this study was to study the transition for Cassie and Baxter regime to Wenzel regime by changing the distance between the pillars. To create patterned Si, two series of nine samples each were fabricated using photolithography.[22] Series 1 has 5-μm diameter and 10-μm height flat-top, cylindrical pillars with different pitch values (7, 7.5, 10, 12.5, 25, 37.5, 45, 60, and 75 μm), and Series 2 has 14-μm diameter and 30-μm height flat-top, cylindrical pillars with different pitch values (21, 23, 26, 35, 70, 105, 126, 168, and 210 μm). The pitch is the spacing between the centers of two adjacent pillars. The Si chosen were initially hydrophilic, so to obtain a sample that is hydrophobic, a self-assembled monolayer (SAM) of 1, 1, -2, 2, - tetrahydroperfluorodecyltrichlorosilane (PF$_3$) was deposited on the sample surfaces using vapor phase deposition technique.[22] PF$_3$ was chosen because of the hydrophobic nature of the surface. The thickness and rms roughness of the SAM of PF$_3$ were 1.8 nm and 0.14 nm, respectively.[23]

The static- and dynamic (advancing and receding) contact angles, a measure of surface hydrophobicity, were measured using a Rame-Hart model 100 contact angle goniometer and water droplets of deionized water. For the measurement of static contact angle, the droplet size should be small but larger than dimension of the structures present on the surfaces. Droplets of about 5 μL in volume (with diameter of a spherical droplet about 2.1 mm) were gently deposited on the substrate using a microsyringe for the static contact angle. The receding contact angle was measured by the removal of water from a DI water sessile drop (~5 μL) using a microsyringe. The advancing contact angle was measured by adding additional water to the sessile drop (~5 μL) using the microsyringe. The contact angle hysteresis was calculated by the difference between the measured advancing and receding contact angles. The tilt angle was measured by a simple stage tilting experiment with the droplets of 5 μL volume.[24] All measurements were made by five different points for each sample at 22 ± 1 °C and 50 ± 5% RH. The measurements were reproducible to within ± 3°.

RESULTS AND DISCUSSION
     Formation of a composite interface is also a multiscale phenomenon, which depends upon relative sizes of the liquid droplet and roughness details. A stable composite interface is essential for the successful design of superhydrophobic surfaces. However, the composite interface is fragile and it may transform into the homogeneous interface. Nosonovsky and Bhushan [25] have studied destabilizing factors for the composite interface and found that the sign of the surface curvature is important, especially in the case of multiscale (hierarchical) roughness. A convex surface (with bumps) leads to a stable interface and high contact angle. Also, they have been suggested the effects of droplet's weight and curvature among the factors which affect the transition.
     Jung and Bhushan [26] developed the model to predict the transition from Cassie and Baxter regime to Wenzel regime based on the factors discussed above. First, they considered a small water droplet suspended on a superhydrophobic surface consisting of a regular array of circular pillars with diameter D, height H, and pitch P. The local deformation for small droplets is governed by surface effects rather than gravity. The curvature of a droplet is governed by the Laplace equation, which relates the pressure inside the droplet to its curvature.[4] The curvature is the same at the top and at the bottom of the droplet.[27,28] For the patterned surface considered here, the maximum droop of the droplet occurs in the center of the square formed by the four pillars as shown in Fig. 3(a). Therefore, the maximum droop of the droplet (δ) in the recessed region can be found in the middle of two pillars which are diagonally across as shown in Fig. 3(a), which is $(\sqrt{2}P - D)^2 / (8R)$. If the droop is much greater than the depth of the cavity,

$$(\sqrt{2}P - D)^2 / R \geq H \qquad (5)$$

then the droplet will just contact the bottom of the cavities between pillars, resulting into the transition from Cassie and Baxter regime to Wenzel regime. Furthermore, in the case of large distances between the pillars, the liquid-air interface can easily be destabilized due to dynamic effects, such as surface waves which are formed at the liquid-air interface due to the gravitational or capillary forces. This leads to the formation of the homogeneous solid-liquid interface.
     To validate the model, contact angle measurements on micropatterned sampled with a range of pitch values were made using droplets of 1 mm in radius (5 μL volume).[24,26] The contact angles on the prepared surfaces are plotted as a function of pitch between the pillars in Fig. 3(b). A dotted line represents the transition criteria range obtained using Eq. (5). The flat Si coated with PF$_3$ showed the static contact angle of 109°. As the pitch increases up to 45 μm of series 1 and 126 μm of series 2, the static contact angle first increases gradually from 152° to 170°. Then, the contact angle starts decreasing sharply. Initial increase with an increase of pitch has to do with more open air space present which increases the propensity of air pocket formation. As predicted from the transition criteria (Eq. 5), the decrease in contact angle at higher pitch values results due to the transition from composite interface to solid-liquid interface. In the series 1, the value predicted from the transition criterion is a

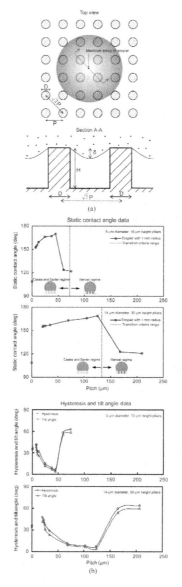

Figure 3. (a) A small water droplet suspended on a superhydrophobic surface consisting of a regular array of circular pillars. The maximum droop of droplet occurs in the center of

square formed by four pillars. The maximum droop of droplet (δ) can be found in the middle of two pillars which are diagonally across, and (b) static contact angle (A dotted line represents the transition criteria range obtained using the model, and (c) hysteresis and tilt angles as a function of geometric parameters for two series of the patterned surfaces with different pitch values for a droplet with 1 mm in radius (5 µL volume). Data at zero pitch correspond to a flat sample.[14,26]

little higher than the experimental observations. However, in the series 2, there is good agreement between the experimental data and the theoretically predicted values for the transition from Cassie and Baxter regime to Wenzel regime.

Figure 3(b) shows hysteresis and tilt angle as a function of pitch between the pillars.[24] The flat Si coated with PF$_3$ showed a hysteresis angle of 34° and tilt angle of 37°. The patterned surfaces with low pitch increase the hysteresis and tilt angles compared to the flat surface due to the effect of sharp edges on the pillars, resulting into pinning.[12] For a droplet moving down on the inclined patterned surfaces, the line of contact of the solid, liquid and air will be pinned at the edge point until it will be able to move, resulting into increasing hysteresis and tilt angles. For various pitch values, hysteresis and tilt angles show the same trends with varying pitch between the pillars. After an initial increase as discussed above, they gradually decrease with increasing pitch (due to reduced number of sharp edges) and show an abrupt minimum in the value which has the highest contact angle. The lowest hysteresis and tilt angles are 5° and 3°, respectively, which were observed on the patterned Si with 45 µm of series 1 and 126 µm of series 2. As discussed earlier, an increase in the pitch value allows the formation of composite interface. At higher pitch values, it is difficult to form the composite interface. The decrease in hysteresis and tilt angles occurs due to formation of composite interface at pitch values raging from 7 µm to 45 µm in series 1 and from 21 µm to 126 µm in series 2. The hysteresis and tilt angles start to increase again due to lack of formation of air pockets at pitch values raging from 60 µm to 75 µm in series 1 and from 168 µm to 210 µm in series 2. These results suggest that the air pocket formation and the reduction of pinning in the patterned surface play an important role for a surface with both low hysteresis and tilt angle. Hence, to create superhydrophobic surfaces, it is important that they are able to form a stable composite interface with air pockets between solid and liquid.

Based on our results of effect of surface structure of patterned superhydrophobic surfaces, such surfaces can be designed for various applications. These can be produced either by lithographic techniques or coating deposition. If one uses the deposition technique, micro/nanostructure (hierarchical structure) can be created by using a combination of micro- and nanoparticles.

CONCLUSIONS

Hydrophobicity, as well as low adhesion and friction, is desirable for many industrial applications. A technique to obtain surfaces that exhibit these types of properties is to study what is found in nature, understand how nature does it, and then mimic it in the lab. Hydrophobic leaves, such as lotus and colocasia, provide perfect samples to learn from and in turn apply these principles in designing superhydrophobic surfaces. Introducing patterned roughness, similar to that found on leaves, is the first step in realizing "biomimetic" surfaces that can be applied to other industrial applications.

We have studied experimentally and theoretically the wetting of patterned Si surfaces with pillars of varying diameter, height and pitch values and deposited with a hydrophobic coating. The experimental observations showed that there is a good agreement between the experimental data and the theoretically predicted transition on patterned surfaces with varying pitch values. In the case of large distances between the pillars, the liquid-air interface can easily be destabilized due to dynamic effects such as surface waves which are formed at the liquid-air interface due to the gravitational or capillary forces. A water droplet in state of the composite interface shows significantly less hysteresis and tilt angles compared to a water droplet in state of the homogeneous interface due to low resistance from the air pockets. The pinning of the droplet at the edges of the pillars takes place during the motion of the droplet and thus pinning results in an increase of hysteresis and tilt angles. These results suggest that the air pocket formation and the absence of pinning in the patterned surface play an important role for a superhydrophobic surface.

REFERENCES

[1]C. W. Extrand, Model for Contact Angle and Hysteresis on Rough and Ultraphobic Surfaces, *Langmuir*, **18**, 7991-99 (2002).

[2]J. Kijlstra, K. Reihs, and A. Klami, Roughness and topology of ultra-hydrophobic surfaces, *Colloids and Surfaces A: Physicochem. Eng. Aspects*, **206**, 521-529 (2002).

[3]Y. C. Jung and B. Bhushan, Contact Angle, Adhesion and Friction Properties of Micro- and Nanopatterned Polymers for Superhydrophobicity, *Nanotechnology*, **17**, 4970-80 (2006).

[4]A. V. Adamson, *Physical Chemistry of Surfaces*, Wiley, NY (1990).

[5]J. N. Israelachvili, *Intermolecular and Surface Forces*, 2nd ed, Academic Press, London (1992).

[6]B. Bhushan, *Principles and Applications of Tribology*, Wiley, New York (1999).

[7]B. Bhushan, *Introduction to Tribology*, Wiley, New York (2002).

[8]B. Bhushan, Adhesion and Stiction: Mechanisms, Measurement Techniques, and Methods for Reduction, *J. Vac. Sci. Technol. B*, **21**, 2262-96 (2003).

[9]B. Bhushan, *Nanotribology and Nanomechanics – An Introduction*, Springer-Verlag, Heidelberg, Germany (2005).

[10]R. N. Wenzel, Resistance of Solid Surfaces to Wetting by Water, *Indust. Eng. Chem.*, **28**, 988-994 (1936).

[11]A. Cassie and S. Baxter, Wetting of Porous Surfaces, *Trans Faraday Soc.*, **40** 546-551 (1944).

[12]M. Nosonovsky and B. Bhushan, Roughness optimization for biomimetic syperhydrophobic surfaces, *Microsyst. Technol.*, **11**, 535-549 (2005).

[13]M. Nosonovsky and B. Bhushan, Stochastic model for metastable wetting of roughness-induced superhydrophobic surfaces, *Microsyst. Technol.*, **12**, 273-281 (2006).

[14]B. Bhushan and Y. C. Jung, Wetting, Adhesion and Friction of Superhydrophobic and Hydrophilic Leaves and Fabricated Micro/nanopatterned Surfaces, *J. Phys.: Condens. Matter*, (2008, in press).

[15]C. Neinhuis and W. Barthlott, Characterization and distribution of water-repellent, self-cleaning plant surfaces, *Annals of Botany*, **79**, 667-677 (1997).

[16]P. Wagner, F. Furstner, W. Barthlott, and C. Neinhuis, Quantitative assessment to the structural basis of water repellency in natural and technical surfaces, *J. Exp. Botany*, **54**, 1295-1303 (2003).

[17]B. Bhushan and Y. C. Jung, Micro- and nanoscale characterization of hydrophobic and hydrophilic leaf surfaces, *Nanotechnology*, **17**, 2758-72 (2006).

[18]Z. Burton and B. Bhushan, Surface Characterization and Adhesion and Friction Properties of Hydrophobic Leaf Surfaces, *Ultramicroscopy*, **106**, 709-716 (2006).

[19]K. Koch, A. Dommisse, and W. Barthlott, Chemistry and Crystal Growth of Plant Wax Tubules of Lotus (Nelumbo nucifera) and Nasturtium (Tropaeolum majus) Leaves on Technical Substrates, *Cryst. Growth Des.*, **6**, 2571-78 (2006).

[20]B. Bhushan, M. Nosonovsky, and Y. C. Jung, Towards optimization of patterned superhydrophobic surfaces, *J. R. Soc. Interface*, **4**, 643-648 (2007).

[21]B. Bhushan, *Springer Handbook of Nanotechnology*, second ed., Springer-Verlag, Heidelberg, Germany (2007).

[22]L. Barbieri, E. Wagner, and P. Hoffmann, Water Wetting Transition Parameters of Perfluorinated Substrates with Periodically Distributed Flat-Top Microscale Obstacles, *Langmuir*, **23**, 1723-34 (2007).

[23]T. Kasai, B. Bhushan, G. Kulik, L. Barbieri, and P. Hoffmann, Micro/nanotribological study of perfluorosilane SAMs for antistiction and low wear, *J. Vac. Sci. Technol. B*, **23**, 995-1003 (2005).
[24]B. Bhushan and Y. C. Jung, Wetting Study of Patterned Surfaces for Superhydrophobicity, *Ultramicroscopy*, **107**, 1033-41 (2007).
[25]M. Nosonovsky and B. Bhushan, Hierarchical roughness makes superhydrophobic states stable, *Microelectronic Eng.*, **84**, 382-386 (2007).
[26]Y. C. Jung and B. Bhushan, Wetting Behavior During Evaporation and Condensation of Water Microdroplets on Superhydrophobic Patterned Surfaces, *J. Microsc.*, (2008, in press)
[27]A. Lafuma and D. Quéré, Superhydrophobic states, *Nature Materials*, **2**, 457-460 (2003).
[28]M. Nosonovsky and B. Bhushan, Lotus Effect: Roughness-Induced Superhydrophobicity, *Applied Scanning Probe Methods* (eds. B. Bhushan and H. Fuchs), Vol. 7, Springer Verlag, pp. 1-40 (2007).

CRITICAL THINKING

David Kalman and Alison Lazenby
Root Learning

OBJECTIVE

Strategy is not being optimized because employees at various levels of the organization do not understand their role in executing it. Our attempts at engaging people in the business trends that shape strategy and the actual execution of that strategy are ineffective. We need a better way.

Something is not working. As leaders, we have refined our vision and mission, our strategic objectives and values until they mean something to each of us and they feel compelling and real. We have debated over our critical success factors and initiatives, our KPIs and balanced scorecards until we have reached a comfortable consensus. We have presented a unified front as we boldly announced our strategic plans and rallied the troops at our annual conferences. We have done all we should, but our best-laid plans are not delivering. Our elegant strategies are not being optimized, and while leaders are left to ask why, the people they lead are asking what more can we possibly do? For critical thinking to thrive, employees need to be engaged in the critical content of the business. In short, they need to be strategically engaged.

Let's begin with what we know:
Strategies are not being optimized.
Strategy execution is the most important factor in delivering results.
People are the key to successfully executing strategies.
People are not engaged or equipped to execute strategies effectively.
The way that we currently engage and equip our people is ineffective.

What is wrong with this picture? Strategies are not being optimized. It comes as no surprise to any architect of a strategic plan that the best-framed plans rarely deliver optimal results. According to Robert Kaplan and David Norton, authors of The Balanced Scorecard, fewer than 10% of well-formulated strategies are successfully implemented.[1] Many remain grudgingly on the paper on which they were written; others find the half-hearted fortune of partial implementation. So, while many of our companies can boast survival from a rocky period, how many of us can say we are truly thriving and delivering the best possible results to our stakeholders?

Top Measures That Matter

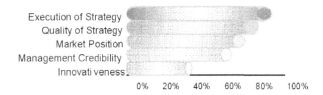

Figure 1: *Source: Ernst & Young*

Strategy execution is the most important factor in delivering results

We have heard about the importance of concise vision statements, focused strategies, core competence,but what really drives sustained business results? Recent studies by Ernst & Young tell us that investors believe the "execution of strategy" is more essential than market position, innovation, or even the quality of strategy itself! We see this play out time and time again. At Root Learning, clients include strategy creators and sponsors from companies across a wide range of industries. Root often finds that its clients' strategies are eerily similar, right down to the metrics, values, and catchphrases each selects after much debate. Larry Bossidy, former CEO of Honeywell, said in his book Execution (co-authored by Ram Charan) that strategy is "no longer an intellectual challenge. You can rent any strategy you want from a consulting firm." While many consulting firms may disagree with this assessment, the meaning could not be more clear. Strategy on paper cannot be the prime differentiator. To truly deliver results and create an environment where people are challenging themselves to find a better way, we must find a way to give strategy a life beyond the paper on which it was written.

However, most leaders devote their energy to the intellectual aspects of the strategy and then delegate its execution to others. This sends a powerful and precarious message. By checking the box once their role in crafting the strategy is "complete," leaders tacitly proclaim that the quality of the strategy trumps its execution. And so, it becomes clearer why 90% of our strategies are not delivering.

People are the key to executing strategies

If strategic execution is the key, then what is the key to strategy execution? Bossidy and Charan state that the "heart of execution lies in three core processes: the people process, the strategy process, and the operations process."[ii] More and more, companies recognize people as the cornerstone of the three. In fact, when Watson Wyatt's Human Capital Index rated companies in managing their human capital performance, they found that "scoring high in 30 key areas of people management relates to about 30 percentage points of return to shareholders."[iii] On the flip side, 70% of all change initiatives fail due to people issues – inability to lead, lack of teamwork, unwillingness to take initiative, inability to deal with change, and other areas.[iv] This makes sense when we think of how technological advancements now allow us to measure operational efficiencies down to the penny and how the combined mind-power of our brightest consultants now ensures that our strategies are first-rate. So, if our strategy and operations are critical arteries, then our people are at the heart of strategy execution.

This means we must connect people in meaningful and integrated ways around our strategies and operations. It means we must translate our strategic ideals to tangible individual accountability, responsibility, and action at all levels of our companies. It means we must reach people beyond pep rallies or intellectual exercises. To deliver sustained results, we must engage employees in ways that are real, meaningful, and compelling to them.

People are not engaged or equipped to execute strategy effectively

We know what we must do. Are we doing it? Put simply, no. In fact, about 71% of employees today do not consider themselves actively engaged in their work.[v] Why does it matter? Disengaged employees are less focused, less effective, and less able to engage customers. In total, this translates to disengaged employees costing the U.S. economy $300 billion per year in lost productivity.[vi] And, this number only figures in the direct cost of lost productivity; imagine how it would grow if we could factor in the opportunity cost of sub-optimizing the most important driver of strategy execution, our people!

The Cost of Disengagement

Feeling unconnected and
uninspired, half of employees
indicate they put in "as little
effort as possible at work."

Figure 2 : *Source: Patterson et al., 2002*

Of course, we all have employees who are engaged in their work, who have all the best intentions to contribute and to deliver. However, many of these employees simply do not know how to deliver according to the expectations placed on them.

In fact, only 43% of employees feel they are given the skills needed to fulfill their job responsibilities and achieve their company's goals.[vii]

To make this worrisome news even more alarming, the complexity of engaging and preparing employees is certain to increase. In the days when companies operated domestically with fairly homogenous staffs, basic employee preparation was well within grasp. Today, globalization, changing workforce demographics, and diversity are making the task of engaging workers more and more complicated. And, the shift to an increasingly services-oriented economy means that we must rely on our employees to make decisions more autonomously than ever before.

So, what are we doing to engage and prepare our employees? Many leading companies are organizing sophisticated learning programs, creating "universities of excellence," and delivering an abundance of talent management programs. This investment in human capital is commendable. But the real bellwether is in evaluating how our efforts translate to business results. So far, the results are murky at best.

The way that we currently engage and equip our people is ineffective

Leading companies are investing significantly in sophisticated training and learning programs. But employees seem to attribute little value to the courses that are so carefully designed. In fact, the majority of workers attribute only 10% of their own job proficiency to formal training – courses and books. [viii]

Yet, we typically see 90% or more of our training resources earmarked for the formal programs, which only deliver 10% of learning. The return on that investment is certainly worth exploring!

Most formal training programs often start and end in the classroom, without any structured opportunity to practice "on the job" where real learning occurs. Is the answer to remove the classroom? Of course not! But, there are two opportunities. The first is to take advantage of the "70%" and simply recognize that informal does not have to mean "random." We have the power to proactively plan and manage the "informal." The second is to redefine the "10%" to make it more effective. This means applying the core concepts of informal learning to the formal setting by engaging employees in strategic context, relevant information, and interactive application.

Further, most formal training courses are geared toward developing hard and soft skills, which are seldom connected and customized to strategic execution plans. Root Learning studies show that this approach is inadequate for two reasons:

• First is the great divide between learning and business where learning specialists are fairly disconnected from business strategists. We see the consequences of this when companies who design comprehensive multi-year strategies do not regularly invest in integrated multi-year learning plans to support those strategies. This divide results in strategies that are not understood and learning that is not strategic. As a result, neither strategy nor learning is as effective as it could and needs to be.

• Next, training courses are traditionally designed as stand-alone, content-based events rather than behavior-based, end-to-end systems. Individual courses are disconnected not only from each other, but also from the bigger strategic picture. Employees are left wondering how the skills they are taught directly contribute to business results or career growth.

If people are the key, what is the key to the people?

• 10% of learning is from formal training.
• 20% of learning is affected through coaches and mentors.
• 70% of learning happens informally through on the-job and off-the-job interactions.

Figure 3: Source: *Lombardo and Eichinger, 2003*

An Alternative

Most organizations have a three-to five-year strategic plan and an operating plan to go with it. When it comes to engaging people in these plans, the effort tends to be event-driven rather than process-driven. The key is to develop an engagement strategy that runs parallel to and in support of the company strategy and the operating plan. An alternative is to develop a long-term, strategic engagement process that enables employees at all levels to share a common mental model around the strategy. Strategic engagement can foster and support an environment where critical thinking is commonplace. Ultimately, this level of engagement translates strategic plans to strategic understanding to strategic action in ways that lead to desired results.

Critical thinking can thrive when we provide employees with opportunities to internalize the critical content of the business in ways that allow them to apply their understanding in complex situations. We can provide employees with:

• Contextual understanding

• Content knowledge

• Opportunities to practice and apply what they learn

Traditionally, companies leapfrog the "Why" and the "What" in order to expedite the decision-making required for the "How." However, without a direct line of sight into how their individual decisions and behaviors contribute to the bigger picture, people's actions will not necessarily be connected to the strategy and cannot be effective in executing it. Let's explore this.

First, by providing contextual understanding of the factors influencing the strategic direction, we can translate what is on paper to clear and consistent understanding at all levels of an organization. Simply said, when it is put into context, strategy can and will make sense.

Once we understand the overarching strategy, it is amazing how it becomes easier to grasp supporting pieces. A well-executed strategy requires many different elements to come together and connect. Typically, employees are asked to put the strategy puzzle pieces together and without ever having seen the entire picture. A cohesive, systemic view makes what once seemed overwhelming and random appear cohesive and sensible. Employees can effectively build their content knowledge of each strategic initiative, skill, or activity once they can place it within the bigger picture. They can begin to translate their understanding to action and the decisions they make.

In order to translate strategic understanding to strategic action, employees must be given the opportunity to practice. Complex decision-making requires sound judgment. Judgment comes through experience. By planning opportunities for trial, error, reflection, and refinement, we accelerate the learning process in a way that is personal, applicable, and sustainable.

Finally, in order to contribute to tangible strategy execution, people need tools and capabilities to apply concepts in real work settings. This happens "on the job" rather than "in the classroom."

Each person in an organization has a role to play in executing strategy and engaging and preparing others to join. It is not only critical that that person excels in their individual role, but also that their efforts are connected to others' and to the whole. However, in most organizations there is often little connectivity or alignment.

Strategy Engagement Process – A closer look

Of course, alignment and consistency do not automatically translate to uniformity. Not everyone in an organization has the same information needs. While it is valuable to have everyone in the organization share a common mental model around the "why" and the "what" of the strategy, "how" it is brought to life will vary according to a person's role. How we expect a CEO to execute her role in the strategy is typically quite different from how we expect a front line worker to contribute. So, in order to bring a strategy engagement process to life, the responsibilities vary based on where one sits in the organization. Let's explore the particular responsibilities unique to each level more closely.

Senior Leaders

Clarity and alignment at the most senior level sets the foundation for how a strategy will be executed. Although tough to acknowledge, it is not unusual to have a wide variance between what is said and what is meant at the senior leadership level. McKinsey Quarterly points to a well-known energy company where five top executives were asked to list the company's 10 highest priorities. Alarmingly, they listed a total of 23 priorities; only two appeared on every executive's list and only seven were on the lists of more than three members. In fact, 13 of the 23 priorities appeared on only one list.[iv] Root Learning's experience is similar. Root's CEO, Jim Haudan, often speaks of the three great lies he hears from senior leaders: "We have a strategy; we are aligned on our strategy; and we have data to support our strategy." If this is true at the most senior level, how then are expectations translated to the broader organization? The messages must be mixed at best.

Create an Aligned Mental Model

Sometimes, leaders are not aligned because our perspectives are simply different. Picture your hometown. What do you see? If six leaders are gathered in a room, we will have six different pictures in our minds. Now, picture "world-class customer service" or "value chain." When we read our strategic documents, we all agree on those priorities. But, what is the picture in each of our minds? Just like our home town, we interpret these words through our individual lenses. While our unaligned interpretation may be completely unintentional, without a common picture, we cannot translate the strategy from paper to action. This is why strategy execution trumps all else. Words say only so much; the translation of the words to meaning and action is difficult. Those who can do it, win.

Cultivate Aligned Behaviors

Sometimes our lack of alignment happens tacitly, but quite intentionally. This happens when our leaders' behaviors do not support the overall strategy through to its full execution, often because team-based rewards compete with corporate results. When this happens, the functional becomes more important than the overall. Leaders' intentions and behaviors then become the core for how strategy is interpreted, prioritized, and executed throughout the functional organizations. The cascade effect can be devastating.

When leaders focus on the creation of strategy and delegate the execution of strategy to others, it becomes challenging to deliver results.

• Instead of entrusting the detail work to others whose time is less "valuable," leaders must adapt their behaviors to truly support strategy execution on par with strategy creation.

• Instead of delegating the "people process" to the once disconnected Human Resources staff, we must bring our most capable HR partners to the strategy table and ensure that people plans are fully integrated with the strategy and operations processes. This requires us to act in an inclusive, people-focused manner, which may demand behavioral shifts or cultural transformation.

When leaders prioritize people as critical drivers of organizational success, we send a powerfully positive message. When we delegate people to the next level, we send a conflicting message to our high-potential managers about what they should focus on to advance their careers to the executive level. This all demands a new degree of clarity, alignment, and accountability among our most senior leaders.

Managers

Perhaps the most untapped source to unleashing momentum and driving change lies in the hands of management. Managers are closest to the people who are working on the front lines every day and have a tremendous ability to directly influence the translation of strategy to action. Perhaps, they are labeled "middle" management not because they are in the mid-stage of their career, but because they are centrally positioned to bridge the gap between senior leaders' strategic plans and individual contributors' daily work. Yet, many managers lack the confidence, experience, and tools necessary to manage effectively.

Interpret Strategy Consistently

This often starts with an inconsistent interpretation of the strategy itself. This may happen because the executive leading a manager's functional area has intentionally or unintentionally shared unaligned perspective and priorities. Or it may happen because the manager simply has not been given opportunity to develop contextual understanding of the strategy. In either case, this inconsistent interpretation leads to unpredictability in executing the strategy.

Connect Team Efforts to Corporate Goals

Creating critical connections so that efforts are aligned to deliver results at all levels is difficult. Often, managers do not have the capability or the supporting tools to help them prioritize activities and make decisions that directly contribute to the overall performance. Too often, decisions that managers make for their teams not only sub-optimize the corporate goals, but actually conflict with other teams' priorities and goals. This leads to disconnected siloed efforts that do not foster optimized corporate results.

Engage People and Teams

Too often, however, this group is unprepared to understand the bigger picture and ill-equipped to engage their employees to perform. Managers have a hard time getting it and have a harder time transferring it to others and translating it to results. This is often the natural consequence of the head violinist being promoted to conductor as a reward for his stellar performance. Individual performance does not automatically translate to group leadership.

Should we expect managers to automatically be prepared to lead upon their promotion? It turns out that the experience they have as individual contributors may not prepare them, and the preparation they receive in the classroom is ineffective. In fact, 73% of graduates surveyed in a recent study report that their MBA skills were used "only marginally or not at all" in their first managerial assignment. The study showed that "one learns to be a leader by serving as a leader."[x] Of course, there must be a more efficient way to prepare leaders than waiting for 20+ years of experience to sink in. To suggest that only the most tenured managers are capable of leading is to miss a critical developmental opportunity. We must find ways to make classroom activities more relevant and to make informal, on-the-job learning more directed. Better yet, we must help managers to be more effective in engaging their people.

Individual Employees

Front line workers are the closest to the customer, to the line, and to the opportunities for tangible change and results. It is through them that the strategy will ultimately live or die. However, about 71% of employees are not engaged in their work and many do not feel they are prepared to do their job.

Understand Organizational and Team Strategies

Like (and often because of) their managers, individual employees' interpretation of strategy is typically incomplete. Often, front line employees have limited information about the organizational direction and priorities. As a result, they may view strategic initiatives as "flavor of the month" because they have not made the critical connections to the bigger picture. We have not presented the strategies in a way that is meaningful and relevant to the individual contributor.

Connect Individual Efforts to Strategic Goals

For many reasons, front line employees don't connect their actions to the whole and don't understand their role in executing strategy. They patiently or impatiently wait for someone to tell them what to do or they vehemently follow what has been outlined for them in their job description. In fact, the percentage of employees who say they have a clear "line of sight" between their jobs and company objectives dropped 13 points between 2000 and 2002 to 52 percent.[xi] In today's fast-paced world, we must ask employees to make decisions at the front line and with customers – and have these decisions align with organizational objectives. We rely on their judgment, but employees are often unclear how to act. Providing these individuals with clarity on how to connect their actions to the bigger picture, along with opportunities to prioritize their personal objectives through performance management systems, will ensure that their actions are executed in the best interest of the organization.

Develop Skills

Once they do understand the strategy and their role in it, it is critical that employees begin to apply that knowledge to their daily work. Employees must translate understanding to action in a way that builds judgment and confidence. Very often, however, skills-training is disconnected from strategic and operational planning. So, employees are not able to prioritize or connect the training on particular skills to the business results or career development opportunity.

Individuals at the front lines are often so entrenched in the demands of their daily work that they are unable to step back and prioritize their efforts or understand the value they contribute in the broader context. And, the training they are provided often exacerbates this situation when it provides formal instruction that is disconnected from strategic priorities. We must find new ways to engage front line employees so they align their efforts with business priorities, they improve their confidence and judgment to make sound decisions, and they commit to developing their careers within the organization.

Engaging People to Deliver Results

We need a better way to engage and equip our people to execute strategy and deliver results.

Deploying strategy is all about creating connections.from the strategic big picture to the tactical daily work, from process change to team performance, and from leaders and managers to front line employees. By building critical connections, people at all levels will understand the drivers of change and will own the strategic response. Teams and individuals can then focus their efforts appropriately. And, our businesses will move strategic plans to focused action and desired results through the broad engagement of our people.

A systems-approach to strategy deployment demands a systems-approach to people engagement. As we focus on the three core processes at the "heart of execution," we create systems for engaging our people that are directly linked to our overall strategy and operations processes. We must integrate learning into our organizations in a way that is strategic, natural, and purposeful. This means creating formal and informal learning opportunities in areas that drive individual performance and business results. It means equipping people with contextual understanding, providing them with relevant and applicable content, and giving the opportunity to apply and practice their new knowledge. And, equally important, it requires engaging an increasingly diverse, global, and sophisticated employee base with creative and engaging methodologies to keep the learning interactive and fun. By combining engaging learning activities with strategic relevance, personal and organizational effectiveness is sure to blossom.

Remember, focusing our smartest leaders on designing sound strategic plans will not set us apart anymore. Exceptional leadership is being redefined. It is about engaging the broader group to execute strategy and deliver results. It doesn't matter how smart we are in the boardrooms, execution is more important than strategy. It doesn't matter what the brightest few know – it only matters how engaged the many are to act and deliver.

REFERENCES

[i] Robert Kaplan and David Norton, *The Strategy-Focused Organization: How Balanced Scorecard Companies Thrive in the New Business Environment* (Cambridge: Harvard Business School Press, 2000).

[ii] Larry Bossidy & Ram Charan, *Execution: The Discipline of Getting Things Done* (New York: Crown Business, 2002) page 22.

[iii] Watson Wyatt Worldwide Human Capital Index: "Linking Human Capital and Shareholder Value."

[iv] Based on research by Ben Zoghi, PhD, Leonard and Valerie Bruce Leadership Chair in Industrial Distribution Program Coordinator & Director of the Thomas and Joan Read Center for Distribution Research and Education, Texas A&M University.

[v] The Gallup Management Journal, June 10, 2004.

[vi] The Gallup Management Journal, 2001.

[vii] Watson Wyatt Worldwide Press Release: "Workers Understand Company Strategy, But Don't Get Skills, Info Needed to Succeed." August 26, 1997.

[viii] Michael Lombardo and Robert Eichinger, *The Leadership Machine* (Lominger Ltd Inc., Otabind edition, December 1, 2000), page 114.

[ix] The McKinsey Quarterly, "Teamwork at the Top," 2001 No. 2 by Erika Herb, Keith Leslie, and Colin Price.

[x] B.M. Bass, *Stogdill's Handbook of Leadership* (New York: Free Press 1981), pages 553-583.

[xi] Watson Wyatt Worldwide. WorkUSA® 2002. "Weathering the Storm: A Study of Employee Attitudes and Opinions."

Additional Information on Figures

Figure 1 - Top Measures That Matter (page 1)
Ernst & Young, "Measures That Matter™ An outside-in perspective on shareholder value recognition" (http://www.ey.com/global/Content.nsf/UK/CF_-_Library_-_MTM).

Figure 2 - Cost of Disengagement (Page 3)
Kerry Patterson, Joseph Grenny, Ron McMillan, and Al Switzler, *Better Than Duct Tape: Dialogue Tools for Getting Results and Getting Along* (Plano, Texas: Pritchett Rummler-Brache, 2000), page 6.

Figure 3 - If People are the Key (Page 3)
Michael Lombardo and Robert Eichinger, *The Leadership Machine* (Lominger Ltd Inc., Otabind edition, December 1, 2000), page 114.

Root Learning is a strategic learning company that offers a blend of formal activities and informal tools designed to drive employee engagement and results. Root Learning is located in Maumee, Ohio, with offices in Chicago, London, England, and Zurich, Switzerland.

GLASS MELTING AT CORNING'S RESEARCH FACILITY

David McEnroe and Josh (Michael) Snyder
Corning Incorporated
Corning, New York USA

ABSTRACT

The Advanced Materials Processing Laboratory (AMPL) is part of Corning's corporate research division which supports glass melting and forming for research, development and manufacturing operations. AMPL was founded back in the 1950's (know as Experimental melting) to provide melting platforms and tools for scientists to innovate new glasses and melting processes. The group has grown over the years and now encompasses glass, glass-ceramics and ceramic materials, therefore the name change to Materials processing. The group's charter is to enable new composition research and to provide smaller scale melting information to help solve melting problems as well as support forming development and small volume manufacturing.

AMPL's capabilities include small scale continuous and periodic (non-continuous) melters, induction melting systems and crucible melting furnaces. Forming operations consist of both in-line forming from the melters and off-line forming such as glass extrusion and redraw. The infrastructure of the laboratory allows insight and understanding of the multiple processes that are involved in inventing new products, from melting the glass to forming a final piece. This allows greater flexibility since it is smaller scale, reducing the time for development and eliminating the need for large tank trials that could be costly if unsuccessful.

This paper will provide a general overview of AMPL's capabilities in terms of glass melting and forming and how some of these tools enable learning of first principles to help solve melting issues.

INTRODUCTION

Corning Incorporated has invented a number of processes and glass materials in their 150 plus years of existence. Some of Corning's major contributions are the ribbon machine for light bulb glass envelop manufacturing, TV tubes, telecommunication fiber, flame-deposition of glass, glass-ceramics, silicone and lately, large flat panel sheets for display applications. These concepts started out in a laboratory environment and then grew through innovation into a manufacturing process. The Advance Materials Processing Laboratory (AMPL) is part of Corning's Science and Technology Division and is where current exploratory work is done for new glass melting and forming processes. AMPL was originally called Experimental Melting but with the addition of Ceramic processing, changed its name. Corning Scientists use AMPL as a laboratory to investigate new glass compositions, develop melting & forming techniques and establish small scale pilot plant operations.

The ability to have a variety of melting platforms has enabled the development of many different types of glasses. Some of these glasses are low melting such as phosphate and chalcogenide glasses, while others require high temperatures as in the case of alumino-silicate and alkali-silicate glasses. The capabilities within AMPL allow glasses to be melted and then formed to obtain process parameter windows. Current manufacturing melting and forming processes can be evaluated on a smaller scale to improve the processes. Another attribute of AMPL is the ability to melt small volumes of glasses for both external and internal customers.

AMPL is broken up into several process areas: crucible melting, larger periodic melters, continuous melting and process development lab along with some specialized forming capabilities. Each area has specific furnaces, support equipment and thermal treatment equipment. A general batch and mixing facility supports all the areas and there are laminar hoods for high purity material handling in the process development lab. In order to meet requests, approximately 450 raw materials are on hand, including a variety of different silica, alumina, boron and other oxide sources.

CAPABILITIES IN AMPL

Crucible Melting:

When compositional work is required, the scientist will request several pounds of glass to be melted in order to fabricate samples for characterization. An electronic submission form has been implemented and the scientist can determine the batch materials, melting parameters and forming instructions which enable technicians to follow the scientist's specifications. A variety of different size crucible melts can be accommodated from several 100 grams up to 10 kg. Once a batch has been submitted, it is weighed out and mixed. Mixing can be done by ball milling, turbula mixing (elliptical rotational movement) or via the v-blender. If particle size reduction is required, Corning has a group that can mill materials to specific PSD.

The majority of the crucible furnaces are glo-bar heated, with an upper temperature of 1650°C. If higher temperatures are required, there are moly-disilicide furnaces with capability of going to 1700°C and 1800°C, along with one gas-oxy fired furnace for 1700°C melting. Several of the furnaces have stirring capability in which a stirrer can be lowered into the crucible, stirred for a period of time and then withdrawn. The crucible melting area keeps a variety of crucibles of different sizes and materials such as fused silica, alumina, zirconia and precious metals.

The majority of crucible melts are done to investigate compositions; therefore, patties of glass are usually formed by simply pouring the crucible onto a table and then cutting the patty up for characterization samples. Most of the forming processes associated with crucible melting are free pours on a table, filling molds or rolling sheet. Corning has a finishing shop which can cut, grind and polish glasses for specific measurement such as beam bending, CTE and numerous other measurements. This allows the scientist a method to evaluate new compositions, modify existing compositions, evaluate raw materials, and make precursor cullet for other types of melting.

Figure 1 Picture of a crucible pour to form a bar of glass.

Periodic Melters:

AMPL has two periodic melters enables larger batch amounts to be melted. The first melter consists of a cylindrical precious metal tank with a 30 kg capacity. This is heated using silicon-carbide glo-bars with an upper melting temperature of 1650°C. A vertical tube is attached to the bottom of the cylinder (down-comer) to allow bottom delivery of the melt and is independently heated. The batch is loaded into the furnace by hand via the stirrer port on top of the melter.

The bottom of the down comer tube has a direct fired tip which can be interchanged with other size tips. An optical stirrer can be inserted into the cylinder for mixing.

Typically, a run consists of preparing a cullet heel to load into the melter in order to fill up the down comer tube to prevent batch from migrating into the tube and causing stones during delivery. Batch is added into the melter starting at 10 kg increments, with decreasing amounts in the final loads. A few hours are usually required to let the melter recover temperature and the batch to melt out. Melts can be run for several hours to several days depending on the experiment and what is being investigated.

The second melter has a 100 kg capacity with a rectangular precious metal tank. Like the 30 kg melter, this tank is heated using silicon-carbide glo-bars with an upper melting temperature of 1650°C. The system incorporates two stirrers at each end of the tank, with one at the fill port and the other over the delivery tube. Batch is loaded through a door in the front of the furnace in proximity to the first stirrer, which is an auger type stirrer that mixes the batch into the melt. The second stirrer is an optical stirrer, placed in a recessed area on the backside of the tank located above the delivery tube. The second stirrer is to cut and break up cord prior to the glass entering the delivery tube. This melter is run similarly as the 30 kg melter with incremental batch additions, but with larger volumes.

Figure 2 - Picture showing the 100 and 30 kg melters

These periodic melters are used mainly for two purposes: to scale up a glass composition from a crucible melt to improve homogeneity in the glass along with providing more glass samples for testing; and to make larger glass volumes for special forming development. When a composition is under development and a crucible melt yields positive results the glass is quite often melted in the 30 or 100 kg melter. Melting a larger volume of glass will often yield a more homogeneous glass. This improved quality of glass is important for obtaining better measurements and more glass is required if the whole gambit of testing (thermal, chemical, optical and mechanical) is done.

AMPL is also involved in special forming process development, so the 30 and 100 kg melters are often used to supply glass for this operation. The melters are placed on a mezzanine to allow various forming equipment to be place under them. The bottom delivery tubes can be modified to enable the use of different orifices which can be used in conjunction with the forming equipment. This enables a variety of forming options to be investigated such as pressing, rolling, sheet draw, tubing draw and casting into molds.

Induction Melting:

Induction melting has been beneficial to many research projects and even used for some small-scale production. Using a precious metal liner that acts as a sesceptor to the induced magnetic field generated by the coil allows fast temperature response times and can obtain temperatures above 1700°C. Current improvements in inductions system have made them much more compact and reliable. The need for an infrastructure of refractory materials surrounding the furnace is not required, which improves glass quality from refractory material contamination and reduces overall space for the melter. The openness of the melter allows for the addition of monitoring equipment along with conditioning equipment if required.

AMPL has two induction systems, a 50 kW and 40 kW, and each system has a 10 kW unit to heat up the downcomer tubes. Different size coils and liners can be used, allowing a range of glass amounts to be melted. The induction units can melt approximately 1 to 16 kg of glass, depending on the experiment. With the fast response time, the induction furnace can be loaded cold and then heated up, taking about fifteen minutes to reach 1500°C when the furnace is fully loaded. Small temperature changes stabilize within minutes, with a +/- 1 degree control capability using a pyrometer to read the temperature.

Usual runs consist of starting up the melter with batch in it and allowing it to reach steady state and then doing batch fills until the required amount of batch has been used. The melt is held at temperature for a period of time to let the batch fully melt down and then a stirrer can be inserted. The stirrer is run for a period of time and then the temperature is reduced for delivery. Run times vary anywhere from hours to days depending on the composition and experiment. Forming work usually consists of casting or pressing glass parts.

Figure 3 - Picture of Induction melter

Continuous Melting:

Along with the periodic melters AMPL has a continuous melter used for research and development projects. The benefit, of the small continuous system is the ability to enable experiments without the upset of production tanks and the discarding of large volume of glass. The ability to modify pull rates or change materials and temperature profiles on a smaller scale is an important aspect of this melter.

The smaller scale melter reduces the time required to do a tank turn-over and observe the impact of the process changes. The other benefit is the fact that the glass is continuously flowing, which eliminates the sometimes unwanted effects of a glass being static in periodic melters. With the emphasis on "green" glass compositions, the smaller continuous melter has played a role in redefining fining packages.

The continuous melter within AMPL is an all-precious-metal tube melter. This melter consists of a metal tube having an inlet at one end and a delivery tube at the opposite end. The tube melter also has a couple of sight tubes and thermocouple ports at different locations along the tube. The tube melter is heated electrically with moly-disilicide elements which allow an upper temperature of 1700°C. A batch feeder system continuously feeds in batch, with pull rates in the range of 10 to 35 pounds/hour.

This melter has been a useful tool for investigating raw materials, preparing glass cullet for frits and fabricating cullet for larger tank start-ups. Having no finer incorporated in the melter can pose a problem, but with control of the pull rate and a fining package in the composition, good quality glass can be made. Most of the forming is rolled glass strips which can be checked for defects and chemistry. Other forming can be done with the tube melter but the space beneath the melter limits larger forming equipment. The tube melter has been run for a day up to several weeks depending on the volume of glass required and the process changes in the experiment.

Figure 5 – Picture of tube melter

Special forming:

In addition to the on-line forming processes mentioned above, AMPL has several off-line forming processes. A tubing tower enables the fabrication of glass tubing using a low viscosity approach. Glass is added to a precious metal crucible with a bottom orifice. A precious metal bell is placed in the bottom orifice opening. The glass is then heated and allowed to flow over the bell, creating the tubing, which is drawn down to the final size. This technique is useful for glasses that tend to devitrify at higher viscosity temperatures.

A hot glass extruder is another process used to fabricate tubing or rods of glass. This process uses a die to make the tubing or rod shape. A glass boule is placed on top of the die and heated; a piston is used to push the glass through the die at high viscosities. The ability to fabricate non-circular tubing or rods of glass is one of the benefits of hot glass extrusion.

A draw tower is another forming tool used AMPL. The addition of the draw tower to our forming capabilities has enabled the ability to draw fiber and to redraw other glass components to obtain attenuated forms. Redraw has also been used to obtain specific intrinsic properties of glasses. For example, polarizing glasses require a molecular component to be stretched in order to induce polarization; this is done by drawing the glass under high tension. Mechanical performance of glasses can also be enhanced through the draw process.

Along with the forming equipment are a number of furnaces for heat treatment of glasses. A variety of annealing ovens are available in many sizes; they are also used for processing glass-ceramic materials and other treatment processes. An acid room is used for cleaning the precious metal components, along with an acid station for chemical treatment of glasses.

Figure 6 – Hot glass extruder

Process Development Lab:

The process development laboratory was initially intended for specialty fiber fabrication by fabricating glasses via traditional melting techniques and not a chemical deposition process. The lab has evolved into a facility where glasses are invented, melting processes are determined and final products are formed. Within the lab are the induction melters along with heat treatment equipment and high purity batching hoods. The lab is environmentally conditioned with HEPA filtration and humidity control. The lab also houses the hot glass extruder and draw tower for redrawing glass which has been previously formed from the melt or extruded. A glove box system is in the lab for work on atmospheric sensitive glasses. Having all this equipment in one lab enables development of new glass products with tight control in all the process stages.

One example of the usefulness of the lab is the development of photonic band gap fiber. A glass composition can be developed along with the melting parameters using the induction melters. Boules of glass, which are fabricated from the induction melt can be extruded into specific tubing geometries.

These extruded tubes are drawn down into smaller components and used to make a preform build. This preform build than can be drawn into fiber.

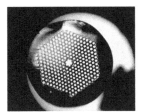

Figure 7 – Picture of photonic band-gab fiber

The advantage of doing all the work in one area with the same people is that everyone understands the entire process and how variations at one stage can impact results down stream. This type of environment enables an awareness of how process modifications up –stream can impact the outcome in the final product.

CONCLUSION

To remain competitive in the current market place, continuous improvement on existing processes is required along with the development of new products. AMPL enables both development of processes and exploratory research of new glasses and glass-ceramics. Providing support of manufacturing operations by having melting platforms to do process improvement without interrupting manufacturing lines has shown to beneficial in both timing and cost. Maintaining state-of-the-art melting and forming equipment provides the researchers access to tools for development of new glasses and also establish both melting and forming processes. This ability speeds up the development cycle for new products. AMPL works with outside companies and academia to provide formed glass in addition to supporting Corning's research and development activities.

Corning's history of glass innovation continues with the next generation of glass and glass-ceramic materials. New industrial applications for glasses are continuing to emerge and areas such as armor, engineered optical materials, substrates, energy devices along with numerous others makes the field of glass science an exciting area.

SMALL SCALE MELTING PLATFORMS FOR PROBLEM SOLVING

M. J. Snyder and David McEnroe
Corning Incorporated
Corning, NY, USA

ABSTRACT:

Small scale continuous and periodic melting platforms can be used as a low cost way to solve problems observed in glass melting. They offer the ability to look at glass systems on a small scale to screen a process for problems which could be encountered on large scale production. A major advantage of these small scale systems is the quick response and turnaround time available to the user.

With small scale continuous melters, short residence times yield fast responses to changes in the system. Effects of compositional alterations, raw material changes, temperature alterations, and fill rates can all be observed in a matter of hours versus the traditional large scale melting response times of days. The major limiting factor of these smaller systems is the short fining times encountered in the melter.

Periodic melter systems offer great flexibility. They have few glass compatibility issues, short preparation times, unlimited melt/fining times, and a large range of temperature capabilities. The major drawback in periodic systems is they must be filled and drained, limiting the ability to analyze continual changes within a system.

The Advanced Material Processing Laboratory (AMPL) at Corning Incorporated has become proficient in small scale glass melting systems. This has enabled significant glass research to address potential melting issues to ultimately improve plant capability and productivity.

INTRODUCTION:

Experimenting with large glass melting systems can take a lot of planning and come with a significant cost. The ability to test and prove an idea in a less risky atmosphere can prove to be a valuable tool in the melting cycle. The Advanced Material Processing Lab (AMPL) at Corning Incorporated has been designed to melt glasses in small to moderate quantities. The melting facilities in AMPL allow scientists, plants, and other organizations to utilize lab scale melting.

Often melting is done in three different platforms:
1. Crucible Melting
2. Small Scale Periodic Melting
3. Small Scale Continuous Melting

These platforms allow an individual/group to pinpoint problems or desirable characteristics in a lower cost/lower risk atmosphere.

The advantages of large scale melting are very evident in the fact that what you see is actually what you get. Typically large scale systems are being run in familiar areas so production can be maintained with little backlash to a production line.

On the other hand, when a problem arises or new compositions need to be run in the melting system, several risks may be encountered. Slow resident times restrict the amount of "new" data obtained from the system. With little feedback, knowledge of what is going on can be detrimental to the whole melting cycle. In addition, long preparation times can be needed, forcing a change to be done at inopportune times.

Due to some of the major disadvantages of changing a process on large scale melting systems, new compositions and process trouble shooting are often avoided or disregarded altogether. Small scale melting can be used to engage in more improvements to the melting systems, ultimately increasing productivity and output quality. The lower cost and time required by these smaller systems allows research to be done with little or no negative impact to the problem or improvement being addressed.

AMPL MELTING

Periodic Melting:

Crucible melting is can be used to obtain a desired compositional range. When the melts are poured, it is easy to make samples for analysis, check for glass impurities, and analyze overall melting efficiency of the glass. After a series of similar glass compositions are melted, the best one from the series can be pinpointed and further analyzed.

In addition to crucible melting glasses can be melted in a larger (10-200 lbs) melter. These melters are basically a crucible surrounded by silicon carbide elements with a hole in the bottom attached to a down comer. A stir rod can be submerged in the melter to thoroughly mix the melt. From here the glass can be delivered in a controlled manner, paying specific attention to the viscosity of the glass. Several forming techniques can be implemented from this melter such as square patties, round/square boules, crushed cullet etc. The advantage of these periodic melters over crucible melting is that very homogenous glass can be formed which allows for further testing on a particular glass.

Small Scale Continuous Melting:

A 10 liter overflow tube melting system commonly used by AMPL lends itself to quick and simple experimentation. Resident times in this melter are measured in hours, allowing changes to the system to be observed very quickly. Typically, the melter is run with glass pull rates in the 10 lbs/hr to 35 lbs/hr range, depending on glass composition. The overflow system allows the melter to be run with a large range of viscosities to satisfy the type of samples being formed during a particular run. Little preparation time is needed and any number of glass compositions can be run in the melter back to back with a short (<1 day) cleaning cycle in between glasses with significant differences in composition. The one major drawback of this system is the fact that the initial melting area, fining area, and delivery system are all one unit. Despite this being a drawback, it yields the compact nature of the melter which gives it the short resident times with efficient continuous melting.

Experiments involving phosphate glasses, lead-silicate glasses, glass ceramics, and alumino-silicate type glasses have all been successfully run through this simple tube melter. In particular, many raw material evaluations have been conducted to compare the melting efficiency of different types of raw materials.

When comparing raw materials, it is important to have a dedicated plan of changes to a certain glass composition regime whether it is the type of raw material or size of the raw material. Several conditions (changes to the raw materials) are created to allow a difference in materials to be observed. A series of batches are arranged in a specific condition order (usually according to a numbering system electronically generated after a substitution is made) to allow for the best transition of changes to be observed. A fill batch is required to fill the melter to the overflow height, usually over a 6 to 8 hour period.

After the fill batch has been added to the melter (using a loss of weight feeding system), the first condition to analyze (typically the same composition as the fill batch) is added to the melter, which causes flow to occur. As stated earlier, batches can be added at a rate of 10 lbs/hr to 35 lbs/hr, where a glass flow rate of 8 – 20 lbs/hr is usually desired according to best results found in past runs. For instance, when using a 15 lbs/hr flow rate, a turnover time of 4 hours is observed in the melter. To allow the current condition to stabilize, 2-4 turnovers are typically used. More turnovers can be used to solidify results, but they are not necessary.

Once the original condition has run its course, a new condition/composition can be added to the melter the same way the prior condition was added after the fill batch. It is important to note that no time in between conditions is needed. For a smooth transition the next condition is loaded in the feeder on top of the prior condition. This can continue for several glass condition changes. Samples are taken at specific intervals to check glass composition or melting nature. Changes to the composition can be accompanied by changes in temperature, which can be done in a ramping mode or with systematic set point change.

Rollers are placed under the melter to form the glass into thin ribbon as seen in Figure 1. The ribbon is typically 3-4 cm wide by 2-5 mm thick. This makes good samples to use for further analysis. In addition to rolled ribbon, other samples can be formed such as crushed dry gauge (sometimes produced to help start larger tank trials), boules, plates, or patties.

Figure 1. Rolled ribbon formed under the AMPL tube melter.

Once all of the desired conditions have been run (several conditions are normally run in a 24 hr cycle for 2 to 5 days for individual trials), a glass matrix of raw material changes vs. defects can be created. This allows an observer to make better informed decisions about specific glasses. Many parameters would be analyzed in a time saving, low cost environment.

An example of a glass matrix that can be formed using the tube melter is shown in Figure. 2. It is a design of experiment evaluating glass fining at a given pull rate versus temperature matrix. In this four box matrix, it is shown that glasses melted at lower temperatures need significantly slower pull rates to those melted at higher temperatures to obtain similar quality glasses.

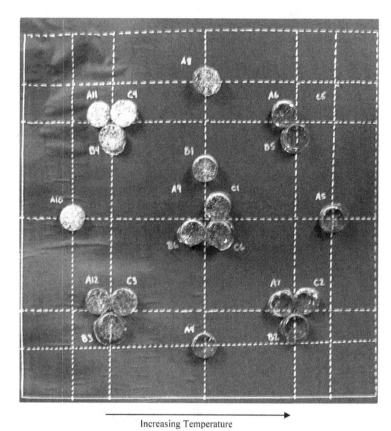

Increasing
Pull rate

Increasing Temperature

Figure 2. Example of Glass fining experiments within a design of experiment matrix

Gas Fining experiments:

In addition to basic glass melting in AMPL, many new melting techniques are tested. An example of testing a new melting technique is glass fining experiments preformed to reduce the uses of toxic fining agents in glass melting.

As the requirement for quality of glasses become more demanding, the tolerance for blisters and seeds has decreased dramatically, both in the number density permitted and the maximum size of the blister/seed. Historically, a variety of chemicals have been added to the glass to help remove the blisters. The most effective of these fining agents, such as arsenic oxide or antimony oxide, are toxic to the environment. The desire to find a more environmentally-friendly method to remove blisters from the glass has led the glass industry to consider a variety of processing techniques. Among these methods is bubbling gases through the melt, a technique known as gas-injection fining.[1,2] Experiments at Corning, Incorporated have demonstrated the use of gas-injection fining to decrease blisters by two orders of magnitude.

Blisters rise in the glass melt at a rate V_s, according to Stokes' Law:

$$V_s = \frac{2}{9} \frac{gr^2 \Delta\rho}{\eta}$$

where g is the gravitational acceleration, r is the blister radius, $\Delta\rho$ is the difference in density between the blister and the glass melt, and η is the viscosity of the glass melt. The larger the blister size, the faster it rises out of the melt.

When using gas-injection fining a strict procedure was followed to ensure reproducible results. The batch material was poured into a platinum crucible, which was placed in a liner inside the induction melting unit. A platinum tube could be inserted through a two-piece refractory lid on top of the liner. A cover gas was introduced over the melt by placing the tube between the lid and the surface of the glass melt. The tube could also be lowered into the glass melt to bubble gas into the melt.

Gas pressure and flow were controlled using a low-pressure regulator and a low pressure flowmeter. During the experiment, bubble formation and release could be verified by periodic changes in flow rate, as indicated by the flowmeter. The induction unit top view is shown with inner crucible in it (figure 3) and being heated with the bubbler tube in (figure 4).

Figure 3. Set up for gas-injection fining

Figure 4. Induction unit with bubbler tube

Blister counts were performed with an automated blister count machine. Figure 5 below shows a comparison of a glass fined with Gas A/Gas B bubbling and a glass fined in air on the same thermal schedule without any gas injection.

Figure 5. Vertical cross sections from glass fined with gas bubbling (left) and without gas bubbling (right).

Figure 6 show the number of blisters/cm^3 produced using the different processing conditions. The best condition for blister removal was to bubble a mixture of 80% Gas A/20% Gas B through the glass.

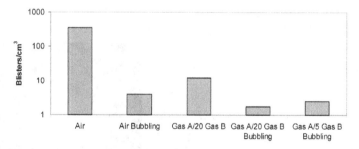

Figure 6. Effect of bubbling and gas composition on blisters

Enabling small size bubbling experiments in the induction unit has provided a tool to evaluate many different experimental parameters and obtain valuable information for fining of glasses. This information can be used to help design experiments for bubbling in different types of melting to help with any related problems which could be encountered. It has been shown that bubbling of a Gas A-Gas B mixture through the melt at a lower temperature helps remove bubbles in the pre-fining step and load high quantity of Gas A into the melt.

CONCLUSION:

The small scale melting platforms at Corning Incorporated can significantly increase chances of developing new glass compositions and help with problem solving on existing glass melting systems. The fact that it is relatively inexpensive while still being effective makes it a great alternative to experimenting on production scale melting systems. In addition to cost reduction, it yields a far less risky environment, allowing for experimenting to be done that would otherwise not be considered.

REFERENCES

[1]I. Peterson, P. Mazumder, S. Schiefelbein, T. Harding, L. Burdick-Burnett, J. Matusick, D. McEnroe, M. Wallen, R. Curreri, and G. Squier, "An Overview of Gas-Injection Fining," *New Glass*, Vol.21, No.4 2006, p18-p24

[2]RGC Beerkens, "Analysis of Advanced and Fast Fining Processes for Glass Melts"; pp 27-30 in Cermaic Transactions, Vol 141 Advances in Fusion and Processing of Glass III (July 2003, Rochester, NY) Edited by J. Varner, T. Seward, H. Schaeffer, American Ceramic Society, Westerville OH 2004.

SOLVING GLASS PROBLEMS

M. G. Mesko, J. T. Fisk
Corning Incorporated
Corning, NY

ABSTRACT

Corning's Materials Engineering group engineers process and product improvements through fundamental materials and process understanding. The group collaborates with and provides a critical link between the research and manufacturing groups. Our strong network results in a broad project portfolio that includes research, development, cost reduction, and quality improvement projects. Changes in laboratory equipment, capability definition, and staffing have resulted in a very efficient, flexible laboratory, which will continue the more than 150 year tradition of Corning Incorporated glass melting test facilities. What makes us successful is that we deliver solutions and recommendations quickly for an ever faster changing market for glass and ceramics products in a cost efficient way. This paper will provide an overview of the changes in our approach to answering glass melting engineering questions.

INTRODUCTION

Corning Incorporated is a world leader in specialty glass and ceramics. It has a 150 year history in materials science, manufacturing, and process engineering. Its product lines include glass substrates for LCD displays, catalytic converter substrates, diesel particulate filters, optical fiber, and specialty glasses that range from Polarcor™ to Macor® and Steuben.

Corning is focused on innovation and discovery and attributes its successes in this area to its unique innovation process. The process brings together research, development, engineering, manufacturing and commercial expertise early in the concept stage.

The Materials Engineering department plays an important role in this process, often providing the linkage between research and manufacturing. The success of the department is a result of its deep technical understanding of materials and processes. In fact, the department has made a concerted effort to focus itself on only a few core competencies. This allows the group to be technically strong and able to apply fundamental principles to a variety of problems. This philosophical change has helped us to become more flexible and effective.

The Materials Engineering department is supported by a world class laboratory where engineers work with technicians to understand new materials, identify defects and how to eliminate them, test material solutions, and more. Focus is also a key theme in the laboratory where only a few different equipment portfolios are maintained, but the diversity and flexibility of the equipment in each portfolio is emphasized. This allows the laboratory itself to be flexible as well.

The intent of this paper is to describe the current focused and flexible approach to addressing glass melting problems through an overview of the laboratory and examples of projects.

VIRTUAL LAB TOUR

The Materials Engineering laboratory occupies approximately 11,000ft^2 of space in the Decker Engineering Building in downtown Corning, NY. The majority of equipment is multi-functional in nature, which allows flexibility in testing as well as fast response to work requests. The lab equipment falls into four major categories:

- Furnaces
- Microscopes
- Analytical tools

- Sample preparation

Each category maintains a broad portfolio of equipment to provide the right tool for any given need.

Furnaces

The lab maintains over 50 furnaces with a maximum temperature capability of 1800°C. Over half of the furnaces are commercially purchased, but a sufficient number of them are designed and built in house. There furnaces vary with regard to their size, materials of construction, heating package, maximum temperature limit, heat up rates, and thermal distribution. This range allows the laboratory to meet almost all of our customers needs. Following is a highlight of a few of our furnaces.

The smallest furnace in the laboratory is a hot stage with a chamber size of 5 cm^3 (Figure 1). The advantage of the hot stage furnace is its low thermal mass, which allows for a fast heat up rate. It is equipped with a fused quartz cover to enable microscopic evaluation of the reactions as the sample is heated.

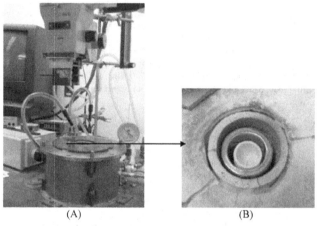

(A)   (B)

Figure 1: (A) Hot stage furnace set-up and (B) 5 cm^3 chamber.

The largest furnace in the lab is a Unique/Pereny furnace. Its usable chamber dimensions are 18 x 18 x 18 inches (Figure 2). This furnace is used to sinter large pieces of ceramic and refractory materials for materials development work and is used for large high temperature simulations.

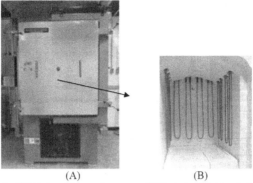

Figure 2: (A) Unique/Pereny furnace and (B) photo of internal chamber.

One of the variables we can control with our selection of furnaces is temperature uniformity. Our gradient tube furnaces are designed to have a 500 – 600°C temperature gradient (Figure 3). These furnaces are probed to determine temperature at 1 inch distances along the entire tube of the furnace. In this way, the sample location within the furnace can be correlated to a temperature (typically +/- 15°C error). These furnaces are useful when a material (or reaction between materials) needs to be evaluated at a range of different temperatures. A platinum boat would be filled with the material, inserted into the furnace for the required time, then removed and evaluated along its length.

When a very narrow temperature range is needed, an isothermal box furnace is used (Figure 3). This furnace has windings on all six sides of the furnace allowing a very tight thermal uniformity of +/- 3°C. This type of furnace is especially useful when studying crystallization properties of glass that are very temperature sensitive.

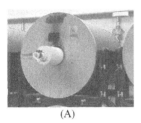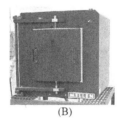

(A)   (B)

Figure 3: (A) Gradient tube furnace and (B) isothermal box furnace

All furnaces in the laboratory are monitored and controlled by a central programmable logic control (PLC) based system (Figure 4). The control function is capable of running simple to elaborate programs based on a combination of temperature, time, and power. The monitor function sends the outputs (temperatures, powers, etc) to a data retrieval system (PI ProcessBook) for easy retrieval and data analysis (Figure 5). Multiple outputs are available from a single furnace if there are multiple temperature zones, or multiple tests which require monitoring.

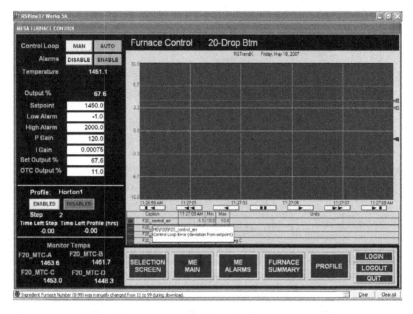

Figure 4: Screen shot of PLC-based furnace control system.

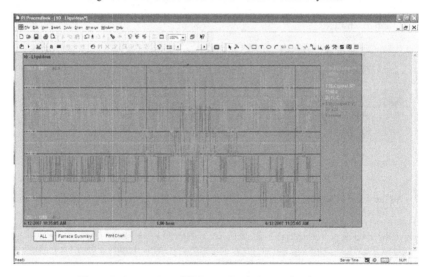

Figure 5: Screen shot of PI ProcessBook data retrieval program.

In summary, the broad portfolio of furnaces in the Materials Engineering laboratory allows for a variety of testing including:
- High temperature simulations
- Material qualifications (refractory and raw materials, as well as glass liquidus measurements)
- Material compatibility
- Heat treatment (ceramic material processing)
- Melting

Microscopy

Microscopy is a critical skill set for any materials engineer. It is used to identify defects and to observe the microstructure of a material, sometimes before and after exposing it to a test condition. For this reason, the Materials Engineering laboratory maintains approximately 20 microscopes of various specifications and capabilities. Even more important than having a range of equipment is having capable staff. The lab is staffed with several members trained in both geology and microscopy and skilled at defect identification and an understanding of phase equilibria to determine defect source.

In the area of optical microscopy, three primary types of microscopes are available:
- Stereobinocular microscopes
- Compound/reflected light microscopes
- Compound/transmitted light microscopes

As with the range of furnaces, each type of microscope provides different capabilities and is used for different applications. Often, more than one scope is used in an analysis to provide deeper understanding.

The stereobinocular microscopes are capable of providing 3-D imaging via two convergent optical paths 12 - 16° apart. Ranges in magnification from 10 to 140X provide resolution down to 5μm. Illumination can occur from any angle, including transmitted and coaxial polarized lighting (PLM). The primary use for this type of scope is to identify raw material contaminants (Figure 6) and to aid in sample preparation.

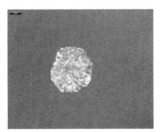

Figure 6: Crystalline quartz impurity found in raw material (bar in upper left corner is 100μm).

The compound reflected light microscopes offer the widest variety of lighting and contrast options including bright field, dark field and polarized lighting. Ranges in magnification from 40 to 1,500X provide resolution down to 0.2μm. The primary use of this type of scope is to observe surface crystallinity, grain structure, grain boundaries, texture, interphases, particulates (Figure 7), scratches, pits, and porosity.

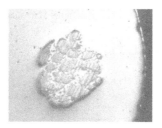

Figure 7: Fused zirconia stone using bright field, reflected light.

The compound transmitted light microscope also offers bright field and polarized light illumination. Magnification ranges from 40 to 1,500X providing a resolution down to 0.1μm. In bright field mode it is used to observe size, shape, color, and index of refraction. In polarized light mode it is used to identify crystalline substance such as minerals, raw materials, refractories, and "stones" in glass product (Figure 8).

Figure 8: Fused zirconia stone using polarized transmitted light (bar in lower right is 20μm).

Several other microscopic techniques can be used to further aid in the identification and understanding of defects. Nomarski differential interference contrast (DIC) imaging is used to enhance the contrast of a surface. It provides higher resolution of surface features based on minor height differences, index of refraction "phases", and reflectivity (Figure 9). It emphasizes scratches and pits on the surface of a sample.

Figure 9: Nomarski DIC of a crystalline polymer coating.

Dispersion staining is another imaging technique used to help identify contaminants, mineral grains, insulation materials (Figure 10), natural and synthetic fibers and virtually any powdered substance.

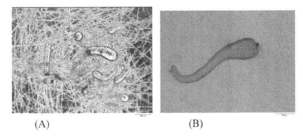

(A)   (B)

Figure 10: (A) Fiberfrax material (bar in lower right is 200µm) and (B) a fiberfrax defect found in glass (bar in lower right is 100µm) obtained by dispersion staining and polarized light, respectively.

The majority of the microscopes in the lab are equipped with digital imaging capabilities not only to provide photo documentation, but to provide input for image analysis if necessary. Image-Pro[k] software is used to provide quantitative measurements of images such as percent area per phase type relationships like porosity, particle size, and phase "a" versus "b". Figure 11 shows an example of the size distribution of particulates on the surface of a sample.

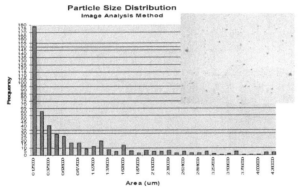

Figure 11: Particle size distribution determined by image analysis.

Image analysis is also useful for accurate measurement of dimensions and geometries. The software is not limited to microscope derived images; it has also been used to provide quantitative analysis of images from inside glass tanks obtained from the department's high temperature periscope (Figure 12).

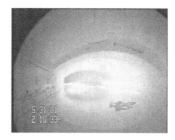

Figure 12: High temperature periscope image of the inside of a furnace. Digital photo may be analyzed using Image Pro software for dimensions and geometry.

In addition to standard fixed optical microscopes, the lab is also equipped with a microscope on an x, y, z stage hooked to an image analysis system allowing automated scanning of samples for gaseous and solid inclusions. The program used to evaluate the image provides number, size, location, and type of inclusions (Figure 13).

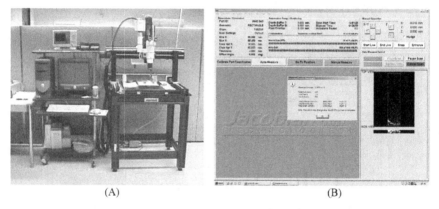

(A)   (B)

Figure 13: (A) Automated inclusion counting system, and (B) output showing locations of inclusions in a glass slab.

Analytical Tools

Corning Incorporated maintains a comprehensive analytical lab facility at its Sullivan Park facility. Even so, there are certain tools that have been duplicated at our laboratory because they are identified as being critical to keep in-house. Duplication is essential because the equipment is used on a daily basis or to aid in process development, so immediate response time is needed. These tools include:

- Scanning Electron Microscopes (SEM) with EDX capability
- X-Ray Fluorescence (XRF)
- Fourier Transform Infra-Red spectrometer (FT-IR) with microsampling microscope

These tools are used to aid in identification of defects and material characterization (including quantitative chemical analysis), product development and process improvements, and lab simulation understanding.

Sample Preparation
  Finally, a range of equipment is maintained in the laboratory to prepare samples before and after testing for analysis. Again, this area could potentially be outsourced to a larger sample preparation laboratory in-house or outside, but the majority of work done in the laboratory is non-standard. Often, samples for testing are one-of-a-kind and cannot be lost in the preparation phase. On the analysis end, engineers and technicians often work together to determine what parts of the sample should be analyzed (eg, if anything occurred during testing that should be studied in more depth), how to cut samples to not damage desirable features (eg, sometimes the sample is rough cut first, then areas of interest slowly exposed so as not to damage microstructure), and to track parts of the sample.

  The sample preparation equipment portfolio includes:
- Saws – large and small diamond saws, clipper saw for larger pieces of refractory, Petrothin for thin cross sections for microscopy
- Drills – core drill as well as precision core drill equipped with various sizes of diamond drill bits
- Surface Prep – Surface grinder for dimensional tolerances and flatness, diamond wheels for rough grinding, lapping wheels for polishing to optical finishes

SOLVING GLASS PROBLEMS

Support of New Glass Development
  Corning has developed thousands of commercial glass compositions in its 150 year history. Efficiency and flexibility are key to getting products introduced to the markets faster. The Materials Engineering department plays a critical role in delivering new compositions to market in three key areas:
- Compositional development
- Raw materials selection
- Refractory prequalification

Compositional development
  The goal of the glass researcher is to discover a composition that meets desired properties like thermal expansion, density, and glass transition temperature. One property of particular interest when transitioning a composition to a manufacturing facility is the liquidus temperature. The liquidus temperature is defined as the maximum temperature at which crystals can coexist with the melt. As you cool the glass, it is the temperature where the first crystals start to grow. In terms of glass forming, it is the temperature you want to stay above until the glass is quickly cooled to a rigid state. This temperature is critical for any forming operation.
  Not only is the liquidus temperature important but so is the devitrification phase. Some crystal phases are small and slow forming. If a process temporarily deviates from set points, these crystal phases may not have enough time to produce noticeable defects. If the crystal phase is fast forming and large, deviations from temperature set points may produce defects in the glass. Even worse, if the devitrification phase formed is difficult to dissolve back into the glass, temperatures must be raised in order to eliminate the defect.

As glass researchers are exploring different compositions, they rely on the lab as well as the department expertise in crystal phase equilibria to determine feasibility of scale-up of a defined composition.

Raw Materials Selection

Another role the Materials Engineering department plays with regard to new glass development is selecting the raw materials that will be used in production. Often, researchers use raw materials that are on-hand (historical materials) or materials that can be readily obtained in small quantities from laboratory equipment suppliers.

This can lead to several problems. Historical materials may be interesting or desirable from a scientific standpoint, but may not currently be available. Alternate materials may have very different specifications, leading to different results. Small scale materials may have purity levels that are unnecessarily high resulting in a cost prohibitive commercial batch or inability to locate a supplier that can provide production quantities.

The role of the raw materials engineer is to identify commercially available materials, starting with materials already existing in the plants and to identify alternates that may improve melting and fining behavior in the glass.

Raw materials are evaluated using a comparison melt test. Glasses are melted and the resulting glass boule is evaluated based on the number of seeds and stones present (Figure 14) and compared to baseline levels.

It is always desired to 'under melt' the standard or baseline batch so comparisons can be made. If the standard contains too few seeds or stones improvements may not be easily recognized.

 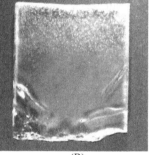

(A)   (B)

Figure 14: (A) Furnace loading for comparison melts and (B) resulting cross sectioned glass boule.

Refractory Prequalification

Refractory testing is required to either specify construction materials for a new glass tank or to prequalify a glass to go into an existing glass tank. These tests are also used if a new refractory material is being qualified. The standard refractory tests include:

• Glass contact refractory and electrode corrosion
• Blister bottom test
• Vapor crown test

The static corrosion test is based on ASTM test C621-84. In short, a finger of glass contact refractory material is suspended in a crucible of glass at melting temperatures typically for 7 days. Analysis of these tests include: evaluation of amount of material lost per unit time, microscopic evaluation of penetration of glass into the refractory, stoning potential, and possibly XRF line scan to look at composition changes across finger thin section (Figure 15).

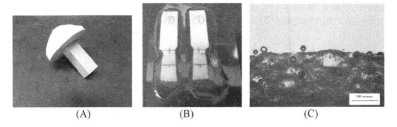

(A)   (B)   (C)

Figure 15: Static corrosion finger (A) before experiment, (B) thin sectioned after experiment, and (C) optical micrograph of dissociation layer.

The dynamic corrosion test is similar to the static corrosion test except the fingers of material are rotated in a crucible of glass. The rotation rate can be based on calculations to simulate glass flow across the refractory material. Analysis of the test is similar to the static corrosion test. Electrode materials can also be evaluated in a similar manner (Figure 16).

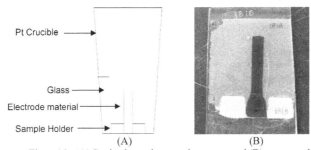

(A)   (B)

Figure16: (A) Static electrode corrosion set up and (B) cross section after testing

Blister bottom test evaluates the propensity of a refractory bottom material to outgas or blister at melting temperatures in contact with molten glass. In this test, a disk of refractory material is placed at the bottom of a crucible, cullet added on top and the glass is held at melting temperature for 7 to 14 days (Figure 17). The glass is then annealed and the number of blisters in the boule is evaluated via microscopy. After blister counts are determined, the glass can be cored and cross sectioned to look at the refractory glass interface for glass penetration, stoning potential, and other potential issues.

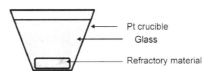

Figure 17: Schematic of refractory blister bottom test.

The vapor crown test evaluates the crown material against batch vapor attack. A sample of refractory material is placed on top of a crucible containing either standard glass batch, or a glass batch with elevated levels of volatile components (to test worst case situations). The test is held for a set period of time prior to cooling. Analysis includes cross sectioning the refractory and observation by microscopy for microstructure changes or XRF line scans for penetration of alkali into the refractory (Figure 18).

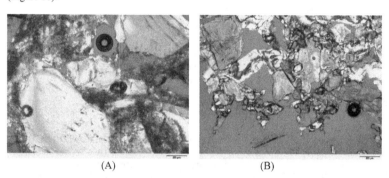

(A)   (B)

Figure 18: Polarized light micrographs of a potential crown material (A) prior to testing and (B) after crown testing. Note this material was not selected as it was evaluated to have a higher stoning potential relative to other materials tested.

Finally, the department plays a key role during tank autopsies. After a tank production campaign is complete or after emergency shut down, the refractory materials and tank design is evaluated (both in situ as well as sampled and evaluated in the laboratory) to look for areas of concern and future improvement.

Manufacturing support
The Materials Engineering laboratory supports manufacturing in two key ways: cost reduction opportunities and quality improvements.

Cost Reduction Opportunities
Many of the cost reduction opportunities involve evaluation of different materials: either raw materials that go into the product or tank refractory materials. In the case of tank refractory materials, specification of new materials is typically done to increase the life of the glass tank to give a lower capital cost per volume of glass. Higher throughput may also be considered a cost reduction as it produces higher volume which means less tanks (capital) need to be utilized.
Material qualification is done in the same way as new composition development and is described above.

Quality Improvement
Defects can come from hundreds of different sources in manufacturing. An inferior paint on a cullet hopper can flake and cause defects. A change in a raw material mine can result in trace contaminants that affect the product. An operations change to eliminate one defect (like changing temperature) may unknowingly lead to another type of defect.

The only key thing in common with these defects and others is the need to eliminate them to maximize production selects. A multi-step technique is used in the Materials Engineering laboratory to provide recommendations to the manufacturing facilities when a quality upset occurs. The technique described below, relies heavily on the staff whose experience in phase equilibria and ability to develop simulations to test hypotheses results in fast understanding and elimination of defects.

Typically the first step in any quality improvement project is identifying and understanding the problem. This is done by accurate and thorough defect characterization utilizing one or more of the microscopes and analytical tools described above. In this step the defect is identified to be either a primary or secondary defect. Primary defects are those whose morphology or chemistry have not been affected by the process. These defects are detected in the finished product in approximately the same morphology as they started. These defects are typically easy to source and therefore, eliminate. Secondary defects are more difficult to source as their morphology or chemistry is different from their starting state. For example, a contaminant may dissolve into a glass at high temperatures then precipitate back out (in a different morphology) at delivery temperatures.

The next step is identification of possible sources. Sources may include raw materials or process materials (eg, tank construction or forming equipment materials). Involvement with the manufacturing site as well as external suppliers is critical in this step to identify changes in these materials.

Next, a hypothesis is developed and tested, usually with a laboratory simulation. Materials are often configured and heated to production temperatures in one of the many furnaces in the laboratory to produce the desired effect. For example if the hypothesis is that a defect is coming from an identified raw material contaminate, the contaminate may be placed in a crucible of defect free glass cullet and heated to production temperatures, held for expected residence times, then cooled and evaluated. If the contaminate remains, the hypothesis may be considered proven. If the contaminate is dissolved into the glass, an alternate hypothesis may be pursued.

Finally, if the hypothesis is confirmed by the laboratory simulation our group communicates with the manufacturing facility and works to help develop an action plan.

CONCLUSION

In summary, the focused, flexible approach that the Materials Engineering department has taken makes it effective at solving glass melting problems quickly. The focus on only a few core competencies allows the department to develop its depth of technical understanding and apply fundamental principles to a wide range of projects ranging from supporting early stage glass development to supporting manufacturing with cost reduction and quality improvement projects. The broadly maintained portfolios in furnaces, microscopy, analytical tools and sample preparation allow the laboratory to be flexible and efficient. Corning's Materials Engineering department has been key to Corning's success over the years and has positioned itself to continue delivering results for years to come.

NUMERICAL SIMULATION OF THE SUBMERGED COMBUSTION MELTING PROCESS

Bruno A. Purnode
Owens Corning Science & Technology Center
Granville, OH

Raj Venuturumilli, Jaydeep Kulkarni
Ansys/Fluent Inc.
Evanston, IL

Lewis Collins
Ansys/Fluent Inc.
Lebanon, NH

Grigory Aronchik
Gas Technology Institute
Des Plaines, IL

ABSTRACT
There is a strong trend in the glass industry to find ways to lower capital and energy costs. The submerged melting technology offers the potential to meet this goal. Submerged melting is a process for producing melts in which fuel and oxidant are fired directly into the bath of material being melted. The combustion gases bubble through the bath, creating an intense heat transfer exchange. The forced convection-driven shear stresses provide rapid particle dissolution and enhance temperature uniformity in the bath.

At this stage, very little is known about the details of the process and very little data exists. Therefore, in order to support submerged melter designs and to better understand the complicated melting phenomena, an important modeling effort has been initiated.

The presence of extremely complex physics and the very disparate times scales between gases and glass flows make solving of the full problem impractical. Therefore a pragmatic modeling strategy has been established in order to find a compromise between faithfully describing the process physics while still relying on reasonable computational cost. The focus of this paper is to present the modeling strategy and the simulation results. An initial comparison of the simulation results with the measured data will also be presented.

INTRODUCTION

The glass industry is considered amongst the top highly energy-intensive industries in the US according to studies conducted in the late 90's by the Office of Industrial Technologies of the United States Department of Energy. The glass industry produces 21 million tons of consumer goods annually valued at $28 billion consuming 400 trillion BTUs, which accounts for approximately 15 percent of production costs. Theoretically, glass making requires about 2.2 MBtu of energy per ton of glass, but usually more than twice that amount is actually used due to various losses.

The Glass Melting Technology Technical and Economic Assessment[1] published by the Glass Manufacturing Industry Council (2006) calls for technological improvements in several areas. The most important is clearly that the glass industry needs a less capital intensive, lower energy cost, and cleaner way to melt glass. It seems like incremental changes in current melters is insufficient for future survival of the industry. Only a Next Generation Melting System (NGMS) using a 'segmented' approach is capable of meeting the Roadmap's High Priority Research Needs. The submerged combustion melting (SCM) technology is the ideal melting and homogenization stage of NGMS. This is a melting approach that provides large capital and energy savings to the glass industry. Submerged combustion melting is a process for producing mineral melts in which fuel and oxidant are fired directly into the bath of material being melted. The combustion gases bubble through the bath, creating a high heat transfer rate to the bath material and turbulent mixing. Batch handling systems can be simple and inexpensive because the melter is tolerant of a wide range in batch and cullet size, can accept multiple feeds, and does not require perfect feed blending.

The need for a new glass melting technology has led to the formation of an unprecedented consortium of five glass companies that are working with the Gas Technology Institute(GTI) to design, demonstrate, and validate the melting stage of a Next Generation Melting System (NGMS). The project is funded by the U.S DOE.

SCM was developed by the Gas Institute (GI) of the National Academy of Sciences of Ukraine and was commercialized a decade ago for mineral wool production in Ukraine and Belarus. Five 75 ton/day melters are in operation[2].

Despite those various industrial applications, very little is currently known about the complex physics present in the process and very little process data exist. In order to support the design of submerged melter and to better understand the complicated melting phenomena, an important modeling effort has been initiated. Therefore a major part of the SCM project is dedicated to this task.

The ultimate goal of mathematical and CFD modeling of fluid flow and heat transfer in submerged melters is to obtain recommendations on their designs and operating parameter. Unfortunately, the presence of extremely complex physics and the very disparate times scales make solving of the full problem impractical. Therefore a specific, stepwise modeling approach has been established in order to find a compromise between faithfully describing the physics of the process while still relying on reasonable computational cost. In this paper, we will present this modeling approach together with comparisons with experimental data.

SUBMERGED COMBUSTION

Submerged combustion melter is a bubbling bath furnace where Fuel and oxidant are fired directly into the bath of material being melted. The high temperature bubbling combustion inside the melt creates a complex gas-liquid structure and a large heat transfer surface. This intensifies the exchange of heat between the products of combustion and the processed materials (see Figure 1 [3]).

The modeling of such a process with several burners is challenging for at least three reasons: first, due to the complex character of the three dimensional physics and chemistry involved in the process; second, due to the very disparate time scales present in the system; and third, due to the coexistence of fast moving turbulent combustion gases and slow moving laminar glass and/or raw materials. Furthermore, it is rather a new process with very little guidance from previous data available from the facilities that run such a process. It is intended in this paper to describe the staged simulation approach employed in this work and discuss the obtained results.

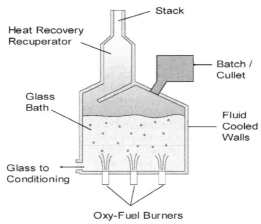

Figure 1: Submerged Combustion.

STAGED MODELING APPROACH

In order to handle the presence of very disparate time scales in the process a multi-staged approach that decouples and distributes the scales among various stages is formulated. This staged analysis strategy is the most promising for balancing accuracy, fidelity to the physics and the associated computational cost (Figure 2).

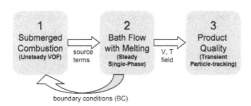

Figure 2: CFD analysis strategy.

The three stages of the analysis are as described below:

1) A 2-D axisymmetric, transient, two–phase (gas/liquid) submerged combustion analysis of a single burner region. The analysis uses the Volume of Fluid (VOF) multiphase method to track the two phases and an eddy-dissipation model for combustion. Turbulence is modeled with the k-ε realizable set of equations. Radiation is modeled using the discrete ordinates (DO) method.

2) 3-D steady-state, single phase analysis of the whole multi-burner melter. This step focuses on the overall flows and heat transfer in the melter and uses equivalent momentum and heat source terms derived from time-averaged VOF results of step 1. Simulations of this step can result in boundary conditions changes to step 1.

3) 3-D transient, tracer species approach, using the velocity and temperature field from step 2) for analyzing the transport and dissolution of the batch by calculating the residence time distribution (RTD) curves.

For complete details about the Volume of Fluid (VOF) method as well as the different combustion, radiation and turbulence models used in step 1 can be found in the Fluent 6.3 reference manual[4]. It must be noted that step 1 is focused on a single burner space with the aim of resolving the combusting bubble dynamics over a time span of several bubbles. To accomplish this we have assumed the region around the burner to be axi-symmetric and strived to situate the outer boundary far enough so that it does not unduly influence the bubble dynamics.

Let us elaborate on the passage from step 1 to step 2. First, the shape of the regions where momentum and heat sources are to be applied is determined by time averaging the phase volume fractions obtained in step 1. The resulting shapes are shown in Figure 4 as glass source regions. Then, flow results from the two-dimensional multiphase combustion analysis (step 1) are averaged and applied as a momentum source $S_{mom,glass}$ in step 2:

$$S_{mom,glass} = C_{mom}(V_{VOF} - V_{fluid})$$ (1)

Where C_{mom} is an arbitrarily large constant, V_{VOF} is the glass phase velocity in the neighborhood of the gas – glass interface calculated from the VOF simulation, and V_{fluid} is the local glass velocity in step 2. Selection of a sufficiently large value for C_{mom} ensures equality of glass velocities between steps 1 and 2 in regions close to the gas – glass interface which itself is not modeled in step 2.

Similarly, heat transfer results from the multiphase combustion analysis of step 1 yield, after an averaging procedure, an energy source of the form:

$$S_{enrg,glass} = \frac{C_{enrg}(T_{flame} - T_{fluid})}{V_{src,glass}}$$ (2)

Where T_{flame} is the average flame temperature (from VOF simulation), T_{fluid} is the average fluid temperature. C_{enrg} is directly related to the average heat transfer rate observed in the VOF simulation, $V_{src,glass}$ is the volume of the glass source region. It is estimated from the step 1 calculation as follows,

$$C_{enrg} = \frac{\dot{Q}}{T_{flame} - T_{glass}}$$ (3)

Where $\dot{Q}$ is the rate of heat transfer from gas to glass, and T_{glass} is the glass temperature. At the top glass surface, the effect of escaping combustion gases is included by applying source terms for the gas flow and heat transfer in small regions of volume V. These regions are shown in Figure 4 as gas source regions. Mass Source can be written as:

$$S_{mass,gas} = \frac{\dot{m}_{gas}}{V_{src,gas}} \tag{4}$$

Where $\dot{m}_{gas}$ is the combustion gases mass flow rate. The momentum source is

$$S_{mom,gas} = \frac{\dot{m}_{gas} V_{gas}}{V_{src,gas}} \tag{5}$$

where V_{gas} is the average gas velocity near the top glass surface level from VOF simulation. Finally, the energy source is:

$$S_{enrg,gas} = \frac{\dot{m}_{fuel} LHV_{fuel} - \int\limits_{V_{src,glass}} S_{enrg,glass} dV_{src,glass}}{V_{src,gas}} \tag{6}$$

Where m_{fuel} and LHV_{fuel} are the mass and lower heating value of the fuel, $\int\limits_{V_{src,glass}} S_{enrg,glass} dV_{src,glass}$ is the net heat transferred to glass and $V_{src,gas}$ is the gas source volume. Thus, this equation assumes complete combustion of the fuel in the melter and imparts the energy not absorbed by the glass to the gas plumes escaping through the top of the glass surface in a consistent fashion with the physical reality.

The final stage (stage 3) involves the calculation of residence time characteristics of the melter. In this stage the flow, component species (gas species, batch and glass) and temperature fields are frozen in time and a short pulse of tracer species is introduced into the domain at time t=0. The average concentration of the tracer at the outlet is constantly monitored throughout the simulation time.

From the outlet concentration versus time curve one can recover the RTD curve between E versus θ which are defined as follows,

$$E = \frac{C_{out}}{\int\limits_0^{t_{sim}} C_{out} dt} \tag{7}$$

$$\theta = \frac{t}{\left(V_{glass} \middle/ \dot{V}_{glass} \right)} \tag{8}$$

Where C_{out} is the average outlet tracer species concentration, t and t_{sim} are the instantaneous and total simulation times, Vglass and $\dot{V}_{glass}$ are the volume and volume flow rate of glass. Thus, $\dfrac{V_{glass}}{\dot{V}_{glass}}$ gives the holding time for the melter.

RESULTS

The result of a single burner multiphase combustion simulation with an E-glass is given in Figure 3 at time t= 10 sec. It shows the mass fraction of glass (red) & gases (blue), respectively. After approximately 10 seconds, the patterns of volume fraction become quasi-periodic and the results can then be used for deriving equivalent source terms for both gas and glass in the single phase calculations, as described in the previous section. In those 3-D single phase calculations, the glass height is kept at 0.71 m. The glass and gas source regions resulting from the averaged stage 1 data is shown in Figure 4.

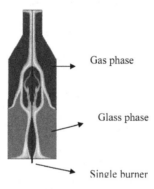

Gas phase

Glass phase

Single burner

Figure 3: Single burner 2-D multiphase simulation.

Temperatures profiles are shown in the plane of the burners are shown in Figure 5. It is clear that we see an increase in temperature from the batch inlet side (right) to the glass outlet side (left). Figure 6 shows the glass temperature in the symmetry plane. Please note that the glass temperature increases and becomes more uniform as we are getting closer to the exit of the melter. This shows the effect of enhanced mixing of the submerged combustion process.

The simulation results of Figure 7 do show that batch is being completely melted in the first half of the melter, showing again the high level of melting efficiency of the submerged combustion process. Preliminary trials in the 1 tonne/hour pilot melter seem to validate the staged modeling approach. Table 1 shows a comparison between the simulation and trial data in terms of glass temperature (at 5" from the downstream set of burners and 4" from the floor on the central plane of the melter) and gas exhaust temperature. The comparison is satisfactory. The instrumented melter allows us to compare the model with a wide range of variables. A complete validation of the model will be done in the coming months as more trials results become available.

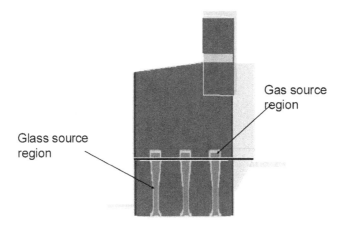

Figure 4: 3-D single phase regions.

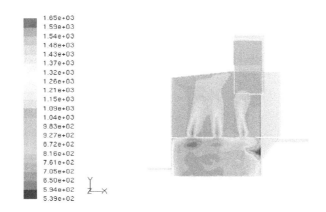

Figure 5: Temperature (K) patterns in the plane of the burners.

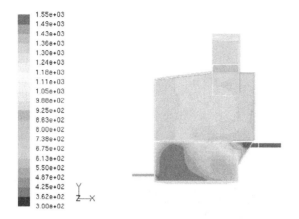

Figure 6: Glass Temperature (K) in the symmetry plane of the melter.

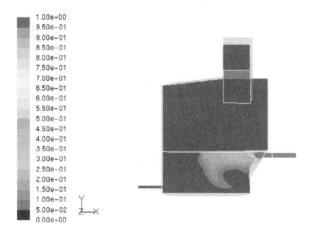

Figure 7: Batch fraction in the submerged melter.

Table 1: Comparison to simulation calculated glass and gas temperature with that of measurements.

Quantity	GTI Trial	Simulation	Difference
Glass T (K)	1649	1532	7%
Flue gas T (K)	1138	1028	9.6%

Figure 8 shows the residence time distribution curve with non-dimensional time on the x-axis and non-dimensional outlet tracer species concentration on the y-axis. It is compared to the perfect mixer, represented by the function $E(\theta) = \exp(-\theta)$. It can be seen from the comparison that the submerged combustion pilot melter produces a residence time distribution that is qualitatively similar to that of the perfectly mixed reactor. It is observed that the outlet tracer species concentration peaks at about 45 minute for the present melter while the holding time is close to 4 hours.

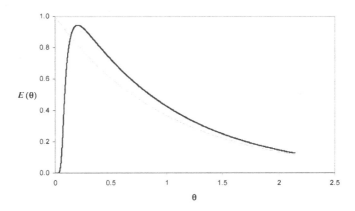

Figure 8: The residence time distribution curve, as compared
to the perfect mixer (dashed line).

CONCLUSIONS

The staged modeling approach for simulating submerged combustion melting established a compromise between accurately describing the complex physics in the melter while still relying on reasonable computational cost. In this approach, results from a transient two dimensional multiphase analysis involving combustion are averaged. This averaging procedure is used in subsequent three dimensional single phase computations. Mathematical modeling of the submerged combustion is an integral part of the whole SCM project. It has been used extensively for designing the pilot melter and its peripherals. The very first trials the pilot melter at the Gas Technology Institute (GTI) seem to support the modeling approach exposed here and seem to validate its results.

ACKNOWLEDGEMENTS

The authors wishes to acknowledge the help of other consortium members: Aaron Huber (Johns Manville), Nic Leblond (Corning) and Carsten Weinhold (Schott) and the financial support of the US Department of Energy.

REFERENCES

[1] Glass Melting Technology, A Technical and Economic Assessment, A project of the Glass Manufacturing Industry Council (GMIC), U.S. Department fo Energy Industrial Technologies Program, Contract #DE-FC36-021D14315, (2004).
[2] Pioro, L. S., Pioro, I. L., Kostyuk, T. O. and Soroka, B. S., in Industrial Application of Submerged Combustion Melters (Kyiv Fact Publishers, 2006).
[3] Olabin, V. M., Pioro, L. S., Maximuk, A. B., Kinkhis, M. and Abbasi, H. A., Ceram. Eng. Sci. Porc. 17 (2), p. 84-92 (1996).
[4] Ansys/Fluent 6.3 Documentation & User's Manual, Lebanon NH (2006).

OPERATION OF A PILOT-SCALE SUBMERGED COMBUSTION MELTER

David Rue, Walter Kunc, and Grigory Aronchik
Gas Technology Institute
Des Plaines, IL

A pilot-scale submerged combustion melter with a nominal continuous pull rate up to 1 ton/h has been designed, assembled, and operated. Tests with E glass, alumina borosilicate glass, and soda-lime glass melts have evaluated impacts of operating parameters such as firing rate, firing pattern, melt temperature, and pull rate on glass homogeneity, volatilization, and other important glass properties. Results are presented along with design philosophy and technical approach for this innovative glass melting system. Comparisons are made between testing results and CFD modeling predictions. Fluent CFD software has been updated specifically to carry out simulations of the SCM unit.

Background

The submerged combustion melter is a bubbling type furnace capable of producing glass and vitreous melts from a number of materials (geologic rocks, sand, limestone, slag, ash, waste solids, etc.). In the submerged combustion melter fuel and oxidant are fired directly into the bath of material being melted from burners attached to the bottom of the melt chamber[1]. High-temperature bubbling combustion inside the melt creates complex gas-liquid interaction and a large heat transfer surface. This significantly intensifies the heat exchange between the products of combustion and the processed material while lowering the average combustion temperature. The intense mixing of the melt increases the speed of melting, promotes reactant contact and chemical reaction rates, and improves the homogeneity of the glass melt product. Another positive feature of the melter is its ability to handle a relatively non-homogeneous batch material. The size, physical structure, and especially the homogeneity of the batch feed do not require strict control. Batch components can be charged either premixed or separately, continuously or in portions.

In the melt bath, heat exchange between the high-temperature products of combustion and the batch particles primarily occurs through the melt. This process occurs in two steps: 1) heat exchange between the products of combustion and the melt, and 2) heat exchange between the melt and the batch particles. Studies have shown that the first step is controlling when producing melt for the manufacture of mineral wool and when vitrifying wastes. For glass melts, and when silica is >40%)of the batch, the second step (dissolution of SiO_2) is controlling[2].

Pilot-Scale Submerged Combustion Melter

Development of the SCM technology has progressed from the work at the Gas Institute of the National Academy of Sciences of Ukraine including the commercial deployment of five air-gas mineral wool melters, through lab-scale testing using oxygen-gas firing to melt a wide range of glasses and mineral melts[3], to the recent commissioning and testing of oxygen-gas pilot-scale melting of commercial glass compositions.

The lab-scale melter assembled at GTI was scaled to the dimensions of the air-fired mineral wool melters and a production rate up to 500 pounds per hour. Oxygen-gas burners were employed for the first time with the lab-scale unit. Most components except burner controls were not significantly modified. A slate of tests over several years provided the input needed to design the pilot-scale SCM.

This larger unit was designed for a maximum production rate of 2000 pounds per hour. The burners had proved stable and reliable on the lab-scale melter, so six new burners were fabricated to tighter tolerances for the pilot-scale melter. Figure 1 shows that the melter walls were again assembled from water-cooled panels. After the panels were built, they were assembled into the full melt chamber and leak tested. After passing this test, the panels were disassembled, and anchors for the castable refractory were welded to all internal surfaces. Some warping occurred during this process. This was remedied by heat treating the panels to straighten them. Panels were reassembled to form the melt chamber, this time in the melter test cell on the melter stand, and then a layer of high-alumina refractory 1.5 to 2 inches in thickness was cast in place on all internal surfaces.

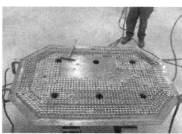

Figure 1. Panel Assembly for Leak Test, Anchor Studs, and Preparation for Refractory Installation

With the melter chamber in place, utilities (gas, oxygen, and natural gas) were connected. The SCM burners operate in a partially premixed mode. The burner control system supplies cold nitrogen during both standard and emergency shutdown to protect the burners and to prevent glass flow into the burners. The combustion system electronics include provisions to set firing rates and oxygen to gas ratios for burners independently or in selected groups. Figure 2 shows an oxy-gas burner, each of which is rated for a maximum firing rate of 1.5 MMBtu per hour. Note that the tip of the burner, which is the only part that contacts molten glass, has been plasma coated with zirconia to protect the burner and to prevent direct metal to molten glass contact. Also shown is the platinum-rhodium discharge tap (not installed) that allows melt flow through the cooled melt chamber wall. This electrically heated tap was specially designed for a research melter to allow test melts with wide ranges of viscosity and melt temperature. The tap is controlled through a power pack and transformer.

Figure 2. An oxy-gas SCM Burner and the Tap Designed for Pilot-Scale SCM Testing

Melt discharge presented a unique challenge. Since tests were designed to test the SCM operation and to allow collection of molten glass samples, a means was needed to safely collect the large, continuous stream of molten glass product. Figure 3 shows the two solutions tried by the project team. The first approach attempted was a conveyor belt installed in a water bath. The molten glass from the tap fell as a steady stream onto the belt and was carried away from the melter under water, cooling as it went, until eventually being lifted and dropped into a collection hopper. This approach worked for several hours but failed because glass particles accumulated in the bottom of the trough under the belt and glass particles would then jam the gears of the belt drive. Despite several efforts to solve these problems that improved performance but did not eliminate the problems, this approach was abandoned. Figure 3 shows the two discharge approaches in operation. The second approach, which ultimately proved workable, was to simply connect a steel trough from the melt discharge position to the top of the glass collection hopper. Glass drops into the trough and is carried downward and cools by contact with water flushed down the trough.

Figure 3. Solutions for Cooling and Safe Collection of Product Glass

Efforts were made to provide as much instrumentation as possible to aid in operation and data collection. Figure 4 summarizes the main instrumentation. Thermocouples measure all gas, oxygen, water, wall, discharge tap, and exhaust gas temperatures. An internal triplex thermocouple in a platinum tube was installed upward through the floor in front of the discharge tap to measure melt temperature. A nuclear gauge was installed to measure average bed level. Gas, oxygen, and water flows are monitored with mass flow meters. Pressures are measured for all gas and oxygen supplies to the burners. For one test a special pressure sensor was installed to monitor burner dynamics at frequencies from 1 to 100,000 Hertz.

Gas and oxygen flow	Into all 6 burners
Gas and oxygen pressure	Into each burner
Water flow	Into each melter panel
Differential pressure	At melter flue exhaust
Melt temperature	Internal 3 TC thermowell
Temperatures	Gas, O_2, water, tap (4 TCs), exhaust gas, refractory
Nuclear level gauge	Average bed height
Digital cameras	Melt surface and tap
Voltage, amps	Tap transformer control
Weights	Batch feed rate

Figure 4. Pilot-Scale SCM Instrumentation and Completed Melter

While the melter itself, including the burners, discharge tap, and sensors, operated flawlessly over the wide range of test conditions selected, the project team had minor problems with peripheral equipment. The Noltec feed system was found to have a narrower range of feed rate than planned, and very fine E glass batch tend ed to bridge in the hoppers. This limited batch feed rates to 1500 pounds per hour. The oxygen system performed well, but the evaporator capacity of 10,000 scfh served as a limit for the alumina borosilicate melt test conducted at a melt temperature of 2950°F. The discharge system for cooling and collecting glass presented problems in early tests, but these difficulties were overcome with the second approach taken to product glass handling. All in all, the pilot-scale SCM unit operated very well. Continuous feed and discharge were easily maintained. Combustion control was reliable and precise. Tests were normally conducted by lighting the burners, filling the bed with melt, testing at continuous conditions that were routinely changed during testing, halted batch charging, and then emptying and shutting down the melter. The time from burner lighting to full melt discharge at steady conditions was 4 to 4.5 hours. Another 4 hours was required after batch feed shut-off to empty the molten glass from the melter, to cool the melter, to clean up, and to secure the unit.

Test Results

Following melter component fabrication, assembly, and shake-down of all systems and instrumentation, the project team met to select a series of glass compositions to melt. A group of four glasses were selected. The chosen glasses are all commercial (although the alumina borosilicate is not currently in production), and they are produced by the team consortium members. Batch for the tests was supplied directly from glass plant batching facilities as premixed batch. The batches covered a wide range of melting properties and glass types. Table I summarizes the four glass batches and their sources. Some glasses contained carbonates and others did not. The precise compositions can not be presented since they are proprietary to the companies shown. The melt temperatures chosen are representative of the temperatures typically used for these glasses under standard practice. Most of the glasses contained boron, but the Al-Ca-Si and the Na-Ca-Si glasses are boron free. The alumina borosilicate glass is alkali-free, which results in a very high melt temperature as shown. To avoid cross-contamination between glass compositions, the melter was thoroughly cleaned between melt tests, including careful removal of the thin, 0.5 inch layer of frozen glass from the melter walls.

Table. I. Pilot-Scale Glasses Melted from April through October 2007

Glass	Comment	Source	Melt Temp F (C)	Tests
Al-Ca-Si	low Na Non-B E glass	Owens Corning	2540 (1395) 2640 (1450)	2
B-Al-Si	Non-Na, K High Si	Corning	2950 (1625)	1
Na-Ca-Si	Soda-lime Glass Windows	PPG	2330 (1275) 2430 (1330)	1
B-Al-Ca-Si	E glass Reinforcement fiber	Owens Corning	2525 (1385)	3

Steady-state conditions for each of the seven pilot-scale SCM tests conducted with the glass batches described in Table I are presented in Table II. Samples were taken in a graphite sample mold throughout the tests and then analyzed by multiple laboratories. Steady-state periods for operation were chosen to be long enough to determine questions regarding volatility of components, product glass composition variations, and energy balances. The first test on 4-04-07 was a shakedown test with a short steady-state period used for final validation of all pilot-scale melter components and instrumentation. This first test was not used for steady-state analyses.

Table II. Pilot-Scale SCM Test Steady-State Parameters

Glass	Al-Ca-Si	Al-Ca-Si	B-Al-Si	Na-Ca-Si	B-Al-Ca-Si	B-Al-Ca-Si	B-Al-Ca-Si
Date	4-4-07	4-24-07	6-25-07	7-19-07	7-26-07	8-30-07	10-10-07
Melt T, °F	2569	2630	2950	2430	2525	2525	2550
Flue T, °F	1728	1850	2325	1700	1580	1775	1800
Pull, lb/h	1572	1122	1212	1084		1200	1200
Bed, lb	4000	4000	5400	3700	3800	4000	4000
Bed, in.	30	30	40	28	29	30	30
St. State, h	0.3*	9	9	6	5	3	5

Exhaust gas sampling was attempted in all tests with limited success. The standard gas analyzers proved unreliable, so the decision was made to switch to 'grab' samples analyzed by gas chromatography. Figure 5 shows GC analyses for gas samples collected during a borosilicate melt test. One sample was collected before batch was fully charged, and the other samples were collected during steady-state operation. Figure 5 shows that the dry gas, as expected, is predominantly carbon dioxide. Small amounts of oxygen and nitrogen are present. The nitrogen is likely from the low concentration present in natural gas. Total NO_x emissions during steady-state operation were measured as slightly above 100 ppmv. When calibrated against the total carbon and the glass pull rate, the NO_x level is found to be between 0.1 and 0.2 lb NO_x per ton of glass during steady state.

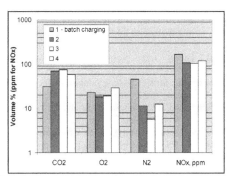

Figure 5. Steady-State Exhaust Gas Composition During Borosilicate Glass Test (8-30-07)

Energy balances at specific pull rates were calculated from steady-state operating data. Data on flows and temperatures of oxygen, gas, water, exhaust gas, and batch were used for the calculation. The energy balances for all four glass compositions are shown in Figure 6. For all glasses except the non-alkali alumina borosilicate melted at high temperature, the results were similar with 34% to 40% energy to glass, 30% to 37% energy loss to the water in the walls, and 22% to 24% of the energy lost to the exhaust gas. The alumina borosilicate glass with a high melt temperature had a lower energy to glass of 27% to 29% of the energy lost to the exhaust gas.

An important consideration for a high-intensity melter such as the SCM is the specific pull rate, here expressed in square feet per ton per day. The lower this value, the more compact the melter becomes. Earlier tests in the lab-scale SCM unit found the specific pull rate to be approximately 1.1 ft^2/ton/day. The recent pilot-scale SCM tests at a production rate of 1500 pounds per hour have found the specific pull rate to be approximately 0.8 ft^2/ton/day. This decrease is a consequence of the larger melter footprint and subsequent decrease in the ratio of wall surface area to melt volume. Figure 6 shows a modeling extension of this decrease to larger industrial melters with an anticipated specific pull rate of 0.5 to 0.6 ft^2/ton/day for 300 and 100 ton/day melters, respectively. Figure 6 also shows predicted energy efficiencies for SCM units of increasing size. The curve was calculated by Fluent CFD and is lower than the efficiencies obtained in the pilot-scale SCM tests. Estimates can be made for larger SCM units based on the pilot-scale data. For an SCM unit producing 120 tons per day of glass, an energy efficiency of 48% to 50% can be anticipated, with no external heat recovery.

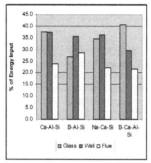 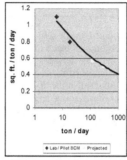 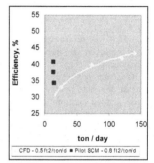

Figure 6. Pilot-Scale SCM Energy Balances, Predicted Specific Pull Rates, and Predicted Energy Efficiencies for Larger Melters

Al-Ca-Si Glass (non-B E Glass)

Major component comparisons for steady-state samples collected during the Al-Ca-Si glass test are shown in Figure 7. Results indicate stable compositions with variations in the major components of less than 3% of their full scale value. For silica, the variation is significantly less than that. Minor components were also found to be stable with no loss by volatility at the temperature of the melter in excess of 2600°F. There was no alkali volatility, which was considered encouraging, but alkali is present in low concentration (under 5%) in this glass. There was no boron in this glass, so later tests were needed to assess boron volatility.

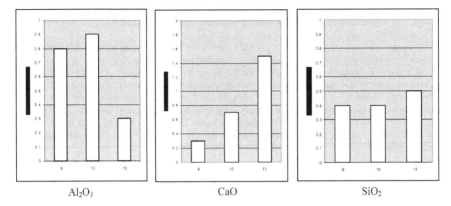

Al$_2$O$_3$ CaO   SiO$_2$

Figure 7. Major Component Variations for Al-Ca-Si Glass During SCM Steady State Melting

Iron redox conditions are important in a number of glasses including the Al-Ca-Si non-boron E glass melted in the SCM unit. Changes in the FeO/Fe$_2$O$_3$ ratio indicate undesired variations in the oxidation conditions inside the melter during melting. Figure 8 shows that the FeO/Fe$_2$O$_3$ ratio was stable during the SCM test with this glass. Some variation is seen, but the Fe content in this glass is low, so additional analyses are needed before determining the actual range of FeO/Fe$_2$O$_3$.

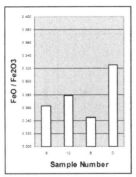

Figure 8. FeO/Fe$_2$O$_3$ Ratio for Al-Ca-Si Glass During SCM Steady State Melting

B-Al-Si Glass (Alkali-free Alumina Borosilicate Glass)

The highest temperature glass melted in the pilot-scale SCM series was an alumina borosilicate glass with no alkali and high silica. This glass batch does contain boron as well as barium. The melter was maintained at a temperature between 2950° and 3000°F throughout steady state melting. The melter itself had no problem with the high temperature. The exhaust gas duct, however, did have problems handling the high temperatures. This duct is lined with high temperature brick, and is thus more susceptible to heat problems than the melter with water-cooled walls. The test was successfully completed with no significant operating problems. Future work with glasses melting at temperatures this high may require additional flue gas dilution with air to prevent overheating.

Steady-state glass samples were analyzed by the GTI laboratory using Atomic Adsorption (AA-ICP) and by the Corning laboratory using XRF. The labs were internally consistent but not consistent with each other. Calibration of analytical methods may be required as shown in Figure 9. Even as the high temperatures of this melt test, no major or minor components were volatilized except for boron. Boron loss was found to be 45% and 35% by the GTI and Corning laboratories, respectively. Overall, samples showed good consistency during steady state operation. The largest components, Al_2O_3 and SiO_2, were found to vary less than 1.5% of their full value. Boron and barium are at lower, but still high, concentrations in this glass and were found to vary less than 8% of their full value in the steady state glass samples.

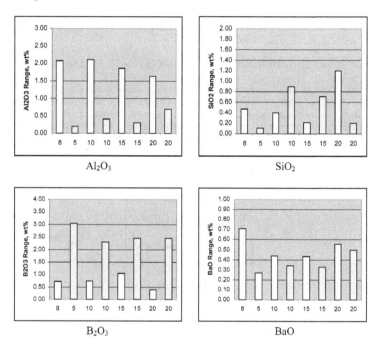

Figure 9. Major Component Variations for B-Al-Si Glass During SCM Steady State Melting

Na-Ca-Si Glass (Soda-Lime Glass)

The soda-lime glass selected for melting was from a PPG plant. The particle size of the batch was larger than the earlier glass batches, and this made charging with the Noltec feed system easier. This glass batch again had no boron, and the melt temperature was significantly lower than for the first two glasses. To obtain a better understanding of glass composition consistency, a larger number of glass samples from the steady-state period were analyzed. Figure 10 shows that the major species were very consistent throughout the steady state period. The three major components varied no more than 0.3% throughout steady state. No volatilization of components, major or minor, was observed during this test. Soda-lime glass is high in sodium, but this often volatile component was found to not be volatilized in the SCM test.

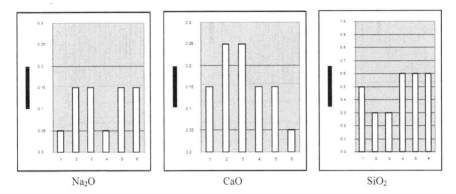

Figure 10. Major Component Variations for Na-Ca-Si Glass During SCM Steady State Melting

B-Al-Ca-Si Glass (E Glass)

The B-Al-Ca-Si E glass has been melted more times in the pilot-scale SCM than any other batch composition. Data, however, is so far only available from the first of these tests. This glass contains boron. Analyses found that boron volatilization was 9%. This is significantly lower than the boron volatilization observed with the B-Al-Si glass and likely reflects the significantly lower glass melt temperature. Figure 11 shows the variation in major component content in the steady-state glass samples. Except for boron, none of the major or minor components of the glass were volatilized. There was, however, more variation in component concentrations compared with the other glasses, with major components varying by up to 7% of their full value.

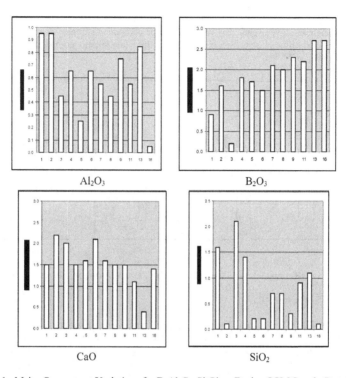

Figure 11. Major Component Variations for B-Al-Ca-Si Glass During SCM Steady State Melting

Further Development

Commercial deployment of a new melting technology is a high-risk undertaking. Glass companies depend on melters to reliably produce glass of needed quality and quantity on a daily basis for many years. For the SCM technology to succeed, a number of development hurdles beyond the present successes must be overcome. The three largest of these are 1) development of rapid and energy-efficient refining to meet the quality requirements of specific glass products, 2) verification of long-term, stable SCM operation, and 3) industrial demonstration of SCM under industrial conditions while producing a commercial product. Work is continuing in each of these areas.

Work is underway to develop a rapid refining process compatible with SCM. If refining is time-consuming and energy-intensive, the gains provided by the high-intensive SCM process will be mitigated. To this end, researchers have first evaluated the refining needs for products such as wool insulation, continuous fiber, container glass, flat glass, and optical glass. This review has shown that some products need little or no refining (wool insulation and sodium silicate in particular), while other products require significant refining. Therefore, the rapid refining process may well be tailored to a particular glass product or segment of the industry. This provides an advantage to many producers since the level of refining employed need only be sufficient for their product.

The refining methods under consideration are being evaluated using high viscosity silicone oil and molten borosilicate glass. The silicone oil provides a low-cost means to test a range of refining approaches at lower cost of time and materials. To date, the methods under consideration have included short residence time (and thin film) refining, ultrasonic refining, and mechanical refining. Results of this work are tentative and not yet available.

Long-term testing or durability validation of the SCM process is planned for future work with specific glass melts. This work is on-going and may be conducted either as part of the rapid refining test program or as proprietary work conducted for individual glass companies. Likewise, the first industrial demonstrations are expected to be for glass companies with needs well met by SCM. These companies will likely need a product with little refining at a scale of 1 to 5 tons per hour. Larger scale operation and the production of glass in industrial facilities is expected to follow this initial demonstration stage.

Acknowledgements

Development of the submerged combustion melting technology has proceeded through the active participation of a consortium of sponsors and technical experts. Funding support has made progress to date possible and is gratefully acknowledged from the U.S. Department of Energy's Industrial Technology Program, the natural gas industry via GRI and SMP programs, the New York State Energy Research and Development Authority, and a consortium of glass companies including Corning Inc., Johns Manville, Owens Corning, PPG Industries, and Schott North America. Technical and in-kind support has been equally critical and has come from the same five glass companies, the Glass Manufacturing Industry Council (GMIC), and Fluent, Inc.

References

1. U.S. Patent 7,273,583, "Process and Apparatus for Uniform Combustion Within a Molten Material".
2. Olabin, V., Maximuk, A., Rue, D., Kunc, W., "Development of Submerged Combustion Technology for Producing Mineral Melts", Int. Gas Research Conf. (IGRC), San Diego, CA, Nov. 1998.
3. Rue, D., Wagner, J., Aronchik, G., "Recent Developments in Submerged Combustion Melting", 67th Conf. on Glass Problems, Columbus, OH, Oct. 2007.

DEVELOPMENT OF AN ADVANCED BATCH/CULLET PREHEATER FOR OXY-FUEL FIRED GLASS FURNACES

H. Kobayashi, E. Evenson, Y. Xue
Praxair, Inc.
Tonawanda, New York

ABSTRACT

Over 10 waste heat recovery systems have been installed on commercial container glass furnaces and have achieved significant fuel reduction. Several of these furnaces are oxy-fuel fired furnaces equipped with batch/cullet preheaters, and they have demonstrated specific fuel consumption of 2.8 to 3 MMBTU/ton for container glass with 50-60% cullet when the batch/cullet was preheated to about 600 to 700 F. This represents energy savings of about 30% relative to state-of-the-art regenerative furnaces. Praxair is currently developing a new advanced batch/cullet preheater system designed specifically for oxy-fuel fired glass furnaces.

A 15 tpd pilot system has been built at the Praxair Technology Center in Tonawanda, NY and preliminary results indicate batch/cullet preheat temperatures approaching 1000F can be achieved with this technology. It is estimated that a commercial version of this system will enable container glass furnaces to operate with projected specific fuel consumption as low as 2.5 MMBTU/ton.

INTRODUCTION

As the glass industry faces rising energy costs, glass furnace operators are increasingly looking to implement technologies to improve energy efficiency. Additionally, in many parts of the world, regulations are being considered to control CO_2 emissions from combustion sources. Oxy-fuel firing of glass furnaces is an established technology that offers the potential to improve furnace efficiency and reduce energy consumption. Actual fuel savings and CO_2 reduction achieved by oxy-fuel conversion depend on the type of the glass furnace and the condition of the heat recovery system in the original air furnace. For large container and float glass furnaces with efficient regenerators, about 10 to 20% fuel savings have typically been achieved. For fiber glass furnaces with metallic recuperators, fuel savings are typically on the order of 40 to 50%. For smaller speciality glass furnaces, which are generally not equipped with efficient regenerators or recuperators, fuel savings of 40-60% are often achieved.

During the last 15 years over 200 glass melting furnaces worldwide have been converted from air to oxy-fuel firing[1]. Oxy-fuel conversion has delivered higher glass productivity and quality, lower emissions of NOx and particulates, lower energy consumption, and lower furnace rebuild costs. Nonetheless, the additional cost of oxygen has created a significant economic barrier for the conversion of many other glass furnaces.

The economics of oxy-fuel conversion would be much more attractive if the energy content of the waste gases could be recovered. This waste heat can be recovered via a variety of technologies as is outlined in Table 1.

Table I – Heat Recovery Options and Fuel Savings for Oxy-Fuel Fired Furnaces
(% savings over oxy-fuel baseline without heat recovery)

Batch Cullet Preheating	Up to 30% (@ approximately 950F)
Oxygen Preheating	5% @ 1000 F
Natural Gas Preheating	3% @ 600 F
Natural Gas Reforming	Up to 20%
Waste Heat Boiler	Steam for Power Generation

Praxair has investigated various heat recovery methods for oxy-fuel fired furnaces since late 1990s. Both oxygen and natural gas preheating were successfully tested at a pilot scale of 1 MMBtu/hr. Oxygen was preheated up to 1200 F in a specially designed tube-and-shell type heat exchanger using hot flue gas from a laboratory furnace. Although care has to be taken to select proper materials for high temperature oxygen service, the basic engineering issues are very similar to the design of a recuperator for hot air burners. Since it is preferred to control the oxygen flow rate of each burner before preheating, a concept of using a cartridge recuperator for each oxy-fuel burner was developed and tested as well. The impacts of oxygen and natural gas preheating on flame characteristics and NOx emissions were also tested to ensure proper burner performance in glass melting applications. Although both oxygen preheating and natural gas preheating were ready for commercial applications, the economics were marginal due to the relatively small fuel/oxygen savings and the low fuel cost (about $3 per MMBtu) in the 1990s. Natural gas reforming recovers thermal energy in the hot flue gas by the endothermic chemical reactions of methane with steam. Natural gas-steam mixture is heated indirectly in catalyst filled tubes placed in a flue gas heat exchanger. Although the potential heat recovery is much greater than the combination of oxygen and natural gas preheating, the complexity of the system and the high capital cost are main drawbacks of the process.

Heye Glas built an oxy-natural gas fired 385 t/day container glass furnace in Obernkirchen, Germany in 1996[2]. The furnace was specially designed with a tall crown to reduce silica refractory corrosion. The furnace is currently in the 11th year of its campaign life with the original silica crown and is scheduled for a cold repair later this year. The furnace has also demonstrated a high heat transfer rate and an excellent energy efficiency with a productivity as high as 2.8 ft^2/tpd (3.5 mtpd/m^2) without electric boost and a specific energy consumption of about 3.2 MMBTU/t with 60% cullet. The furnace is equipped with a waste heat boiler to raise steam and the flue gas is cooled from about 2500 F to about 400 F. The steam from the boiler is used in a steam turbine to generate about 1 MW of electric power. One of the advantages of this heat recovery system is the small volume of the cooled flue gas which is directly fed into a small bag house after sorbent injection for SO$_2$ capture. However, the high capital cost and maintenance requirements for this system have limited widespread adoption.

Of the heat recovery options considered, waste heat recovery via batch/cullet preheating offers the best compromise between potential energy and oxygen cost reduction and capital/maintenance expenditure for oxy-fuel glass furnaces. The integration of a heat recovery system with oxy-fuel firing can provide significant reduction of fuel and oxygen consumption, as well as CO$_2$ emissions, and substantially improves the economics of oxy-fuel conversion[3]. Furnace energy balance analyses show an additional 30% reduction in both fuel and oxygen consumption by combining batch/cullet preheating for an oxy-fuel fired glass furnace.

A number of well documented batch and cullet preheaters have been installed in commercial container glass furnaces over the last twenty years. They include five Interprojekt units[4], three Zippe units[5], two Sorg cullet preheaters and three Praxair (from Edmeston) cullet preheater/ electrified filter units[6]. In these units moving beds of batch/cullet are heated to 500-750 °F by hot flue gas either in direct contact or in indirect contact (Zippe units) and typical fuel savings of 10-20% have been reported. Since batch/cullet preheating also reduces the heating and melting time required within the furnace, the production rate can be increased. Optionally, if high quality is needed, the number of seeds can be reduced due to higher glass melt temperatures. In spite of these benefits, only a small number of furnaces have adopted batch/cullet preheaters due to the relatively large initial capital cost, maintenance and operating costs and concerns over dry batch carryover.

Rapid escalation of fuel prices in recent years and increasing CO_2 credit prices have significantly changed the economics of batch/cullet preheating, especially for oxy-fuel fired furnaces. In the following sections the performance and technical issues reported from the above installations are briefly described. Finally, the development and demonstration of the Praxair advanced batch/cullet preheater, specifically designed for oxy-fuel furnace application, is reported.

FUEL CONSUMPTION FOR OXY-FUEL FIRED FURNACES WITH WASTE HEAT RECOVERY

A number of commercial container glass furnaces have been integrated with waste heat recovery systems and achieved significant fuel reduction. A brief review of the published results of some of these systems is provided below.

Interprojekt Batch Preheater[4,5]

Nienburger Glas (now: REXAM) has the most extensive experiences with batch/cullet preheaters. The first unit was installed in 1987 in Furnace no. 4 and replaced with an improved version in 1999. The Interprojekt system is a direct contact counter-current/cross-flow moving bed heat exchanger. All raw materials are mixed prior to delivery to the preheater. The flue gas downstream of the regenerators travels up through several layers of open-bottom ducts in a counter flow configuration relative to the batch being drawn downward as it is fed into the furnace. Some of the batch particles are entrained into the flue gas and captured in a downstream dust removal unit. At the start of a new furnace campaign, the cross-fired furnaces, end-fired furnace and oxy-fuel fired furnace all show low specific fuel consumptions of about 3.2 to 3.4, 3.4, and 2.9 MMBTU/t[4]. The specific fuel consumption of Nienburger Glas furnace no.4 increased steadily over its twelve-year furnace campaign to about 4.7 MMBTU/t, or about 3.5% average increase per year. Similar annual increases were observed for furnace no.1. The data from later installations show significantly lower annual fuel increases of about 2 % and 1.3%, presumably due to improved furnace/regenerator/preheater system designs. By comparison, the average annual increase in fuel consumption of regenerative furnaces without batch preheaters is reported to be about 1.35% per year in a furnace energy benchmarking study conducted by TNO[7].

An Interprojekt batch/cullet preheater was installed for a new 440 t/day oxy-fuel fired container glass furnace at Gerresheimer Glas (now OI-BSN) in Düsseldorf in 1997. The melter area is 1615 ft² and the specific pull rate is 3.7 ft²/tpd (2.7 mtpd/m²). The specific energy consumption of about 2.8 MMBTU/t was reported with 50 to 70% cullet. Flue gas exits the furnace at about 2550 °F and is cooled to about 1040 °F by injecting cooling air before the batch/cullet preheater. The batch/cullet mixture is heated to about 570 °F and the flue gas is cooled to 355-390 °F. The specific fuel consumption of this oxy-fuel fired furnace with the preheater (GX2-Gerrisheimer Glas no.2) is shown to be relatively constant at about 2.9 MMBTU/t for the first three-year period reported in the paper[4]. (Note: The energy consumption of oxygen generation is not taken into account.)

Zippe Batch/Cullet Preheater[8]

Zippe developed an indirect batch/cullet preheater based on a counter flow-cross flow plate heat exchanger design. This system is designed in a modular form, consisting of individual heat exchanger blocks stacked up vertically. In each module horizontal flue gas passages are separated by parallel steel plates forming vertically passages for gravity fed batch/cullet materials. Hot flue gas criss-crosses the vertical moving beds of batch/cullet materials through the horizontal passages and moves up to the next module above through a connecting side duct. Heat from hot flue gas is transferred indirectly through the steel plates to the moving beds of batch/cullet, minimizing pressure losses of the exhaust gases and avoiding fine particulates entrainment problems of direct contact batch/cullet preheaters. Up to 50% batch mixture (i.e., at least 50 % cullet is required) can be heated by providing special steam venting funnels on top of the preheater in the batch/cullet passages to remove steam and other vapours.

Three commercial units have been installed for container glass furnaces. Batch and cullet mixture is heated to 525-610 °F while the flue gas cools down from 930 °F to 375-390 °F.

Oxy-Fuel Fired Furnace with Praxair (from Edmeston) Cullet Preheater/Filter[6,9,10]

Edmeston AB developed a hybrid cullet preheater/filter system using the electrostatic precipitator principle to capture dust particulates on the moving bed of cullet. The system consists of an ionizer to impart electrical charge to the dust particles in the hot flue gas and a filter module which has a moving bed of cullet with a built-in high-voltage electrode. Three commercial systems were installed. The first unit was installed at the Irish Glass Bottle Co. in 1994. The second unit was installed at Leone Industries in New Jersey, USA for a new 275 t/day oxy-fuel fired flint container glass furnace in 1998[11]. The furnace was equipped with a Praxair cullet preheater and filter system, consisting of a pyrolyzer, an ionizer and a filter module. The third pyrolyzer module (without the filter module) was installed at Vetro Belsamo.

Hot flue gas from the furnace is tempered to about 570 °F with dilution air and recirculated flue gas and introduced into the preheater/filter module for cullet heating and filtration. Cullet is also heated in a separate pyrolyzer to a higher temperature (up to 930 °F) where organic contaminants in post-consumer cullet are vaporized/pyrolized by hot flue gas tempered. The cooled flue gas with the organic vapors is recirculated and incinerated in the hot flue gas duct to eliminate the odour problem found in the first installation. The average cullet preheat temperature is about 660 °F. The preheated cullet is mixed with batch and the dry preheated batch/cullet mixture is charged into the furnace. The specific energy consumption has been consistently about 3.0 MMBTU/t with 50% cullet when furnace production rate was about 265 t/day and no sign of furnace aging effect on energy consumption has been observed for nine years of operation[10].

DIFFERENCES IN RECOVERABLE WASTE HEAT

The batch/cullet preheaters available today were originally designed for a large volume and relatively low temperature of flue gas from air fired regenerative furnaces. In order to apply these systems to oxy-fuel fired furnaces, dilution air is mixed into the flue gas to reduce the flue gas temperature from about 2640 °F to about 932 to 1100 °F. The dilution of hot flue gas not only increases the volume of the flue gas, and hence the size of the down stream gas handling equipment, but also reduces the amount of recoverable heat substantially. Figure 1 compares the amount of recoverable heat from 450 t/day (410 metric ton/day) container glass furnaces equipped with a batch/cullet preheater for (1) air fired regenerative furnace, (2) oxy-fuel fired furnace with dilution air, and (3) oxy-fuel fired furnace without dilution air. The flue gas temperature after the batch/cullet preheater is assumed to be 428 °F

(220 °C) in all cases. For the air fired regenerative furnace the recoverable heat corresponds to the enthalpy difference between 842 °F (450 °C) (i.e., assumed flue gas temperature after the regenerators) and 428 °F (220 °C), or about 55% of 12.9 MMBTU/hr (13.6 GJ/hr) of waste sensible heat. For the oxy-fuel fired furnace without dilution air, about 88% of 14.8 MMBTU/hr (15.6 GJ/hr) of waste sensible heat is recoverable due to the high flue gas temperature of 2642 °F (1450°C). When dilution air is used with oxy-fuel fired furnace to reduce the flue gas temperature down to 1112 °F (600 °C), flue gas volume is roughly tripled and the recoverable heat is reduced to about 63% of the waste sensible heat.

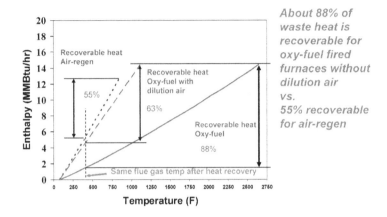

Figure 1- Recoverable waste heat in flue gas – Air vs. Oxy with and without dilution air

Clearly it is advantageous to use the high grade waste heat from an oxy-fuel fired furnace without dilution air. Several options have been previously patented[12]. Recuperators can be used with the hot flue gas to preheat oxygen and/or natural gas and to cool down the flue gas to a temperature acceptable to a batch/cullet preheater. Another option is to reduce flue gas temperature is to install a shadow wall can be installed near the charge end of a furnace. These methods, however, add complexity and costs to the overall heat recovery process and still limit the potential energy recovery.

In order to address this issue, Praxair has developed a batch/cullet heat recovery system that can take the hot flue gas from an oxy-fuel furnace without dilution air. This system minimizes the flue gas volume, simplifies the flue gas handling design and maximizes energy recovery from the waste heat to improve the economics of oxy-fuel fired glass furnace operation.

TECHNICAL CONCERNS WITH BATCH/CULLET PREHEAT SYSTEMS

Dry Batch Carryover

Carryover of dry preheated batch is a major concern for regenerative furnaces, as it would plug the checkers and adversely affect furnace energy performance. The amount of dry batch carry-over into regenerators is influenced by batch ingredients and particle size distributions as well as the gas velocity over the unmelted batch areas. Furnace gas velocities along the path of the regenerator port are typically of the order of 33 ft/s and entrainment of fine particles is common. Actual furnace measurements by TNO of dry and wet batches in a different regenerative furnace showed increased carryover with dry batch, especially when the batch mixture contains fine particulates less than 100 microns in diameter [13].

In oxy-fuel fired furnaces, the momentum of each flame is much lower due to the combination of reduced firing rates and the elimination of ballast nitrogen. The gas velocity near the batch area can be reduced to as low as 3-15 ft/s, especially with the Praxair Tall Crown Furnace design[2] which minimizes the volatilization of alkali species from glass melt and batch surfaces. Thus the amount of dry batch carryover in the flue gas is much lower than that of an equivalent regenerative furnace due to low furnace gas velocities near the batch area.

The experience at Leone Industries[9,10] indicates that batch carryover has not been a significant problem in the long term operation of an oxy-fuel furnace combined with cullet preheating and dry batch charging. There has been no significant accumulation of batch dust in the flue duct and the furnace energy performance has been very consistent over a nine year period. In-furnace measurements of dust particles were conducted recently in two oxy-fuel furnaces with and without batch/cullet preheating at Leone Industries and the measured results confirmed the above observation. Dry batch dust problems can occur in transport lines from the batch preheater to the furnace. if the hot batch conveying system is not properly designed for air tightness. Gerresheimer Glas reported dust problems near the dog house area (outside the furnace) from the vibrating tray hot batch conveying lines[4]. However, conveying systems specifically designed for hot material handling should mitigate these issues.

Batch/Cullet Redox Control

Post consumer cullet contains variable amounts of organic contaminants (food residues, etc.) which influences the redox significantly. Cullet preheating reduces and stabilizes the organic content by de-volatilizing and oxidizing (burning) some of the organic materials in the cullet and batch. In direct contact batch preheaters a significant fraction of carbon in the batch/cullet mixture reacts with the hot flue gas and batch redox has to be adjusted accordingly to compensate for the loss of carbon. Color control problems have been reported for amber glass. This problem would be minimized in an indirect contact preheater.

OTHER BENEFITS OF BATCH/CULLET PREHEATING

Electric boost reduction, roof temperature reduction and production rate increase

The rate of batch/cullet melting becomes faster when the batch/cullet is preheated. If the same furnace production rate is desired, the fuel input and/or electric boosting are reduced as is the crown temperature. Measured results for an end-port regenerative furnace equipped with a Zippe preheater at PLM (now Rexam), Dongen showed a specific fuel consumption and electric boost of 3.7 MMBTU/t and 133 kW respectively without preheating, and 3.4 MMBTU/tGJ/t and 46 kW after batch/cullet preheating to about 525 °F at 360 t/day and with 55-62% cullet [9]. Glass quality (seed count) was improved at the same production rate. Alternatively, the furnace production rate could be increased to take advantage of faster melting with preheated batch/cullet. In addition, the crown temperature was reduced about 40 °F due to the use of preheated batch at PLM and Leone[10].

Reduction of volatilization and particulate emissions

Since the firing rate is reduced at a constant glass production rate, both the batch/glass melt surface temperature and the gas velocity near the surface are reduced at the same pull rate. Thus the reactive volatilization of alkali species, which is the main source of particulate emissions in most glass furnaces, is reduced significantly. Model calculations predict 40-60% reduction in volatilization with batch/cullet preheating[9]. If the physical carryover of the dry preheated batch is controlled by a good oxy-fuel furnace/burner design, a similar reduction in particulate emission is predicted with an indirect batch/cullet preheater.

Reduction of acidic compounds (SO_2, HF and HCl)

In direct-contact batch preheaters, acidic compounds in the flue gas react with batch compounds to form stable salts in the batch and the dust is captured in the electrostatic precipitators or bag filters. The filter dusts are generally recycled back to the furnace. Significant reductions in the emissions of these compounds (56% for SO_2, 62% for HF, and 87% for HCl) were reported at Nienburger Glas[5].

Reduction of selenium addition in flint batch

Volatile compounds (such as selenites used as decolorizing agent) form submicron size particulates in the flue duct or react with batch materials in the batch preheater as they cool down below a certain temperature. The particles are captured in the downstream dust collection system and recycled back to the furnace. Substantial reduction in the selenium requirement was observed both at Nienburger Glas (76% reduction)[5] and at Leone Industries with the Praxair Cullet Preheater/Filter System[6].

ADVANCED OXY-FUEL FIRED FURNACE WITH A NEW BATCH/CULLET PREHEATER

In Figure 2, energy balances for a 450 t/day container glass furnace with 50% cullet are compared for five cases: (1) cross-fired regenerative air-fired furnace, (2) regenerative air-fired furnace with batch/cullet preheater, (3) oxy-fuel fired furnace, and (4) oxy-fuel fired furnace with batch/cullet preheater using dilution air, and (5) oxy-fuel fired furnace with batch/cullet preheater without dilution air. The conversion of the state-of-the-art air-regen furnace (Case 1) to oxy-fuel firing without heat recovery (Case 3) reduces fuel consumption by 13%, from 4.0 MMBTU/t to 3.5 MMBTU/t. With batch/cullet preheating the fuel consumption is reduced by 14.4% to 3.4 MMBTU/t for the air-regen (Case 2), while the fuel consumption for the oxy-fuel with dilution air (Case 4) is reduced by about 30% over the air-regen baseline to 3.0 MMBTU/t. This is consistent with the operating experience at Leone and at OI-BSN. In fact, actual fuel consumption demonstrated at OI-BSN was slightly lower (i.e., 2.8 MMBTU/t) as the cullet ratio was higher (50 to 70%). This operating data validates the energy saving estimates. With a batch/cullet preheater without dilution air the oxy-fuel Case 5 shows that the fuel consumption is reduced by 37% over the air baseline to 2.5 MMBTU/t.

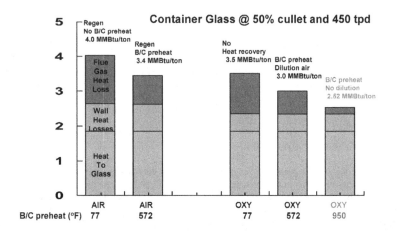

Figure 2: B/C Preheat Temperature vis Furnace Energy Consumption

Figure 3 compares specific fuel consumption of different type container glass furnaces with or without batch/cullet preheaters. The curves for cross-fired and end-port furnaces and the data points for oxy-fuel fired furnaces are taken from the furnace benchmarking study conducted by TNO [7]. Two data points for the Leone Industries and Gerresheimer Glas were added to represent the state-of-the art oxy-fuel furnaces with batch/cullet preheat temperature of about 575 °F. The bottom curve represents the projected specific fuel consumption of oxy-fuel fired furnaces with batch/cullet preheat temperature of 930 °F. The specific fuel consumption is projected to be about 2.5 MMBTU/t for a 330 t/day container glass furnace with 50% cullet. It represents about 30% fuel and oxygen reduction as compared with the direct oxy-fuel fired furnace without heat recovery.

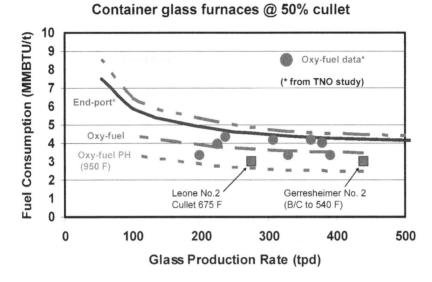

Figure 3: Glass Production Rate vs Fuel Consumption

Praxair is currently developing an advanced batch/cullet preheater for oxy-fuel fired furnaces to preheat batch/cullet mixtures to 900-1100 °F. Previous laboratory tests have shown that the common soda-lime glass batch materials can be heated to these temperatures without stickiness problems. The design objectives of the system include:

- Preheat a wide ratio of batch and cullet simultaneously in the same unit
- Ability to process internal and after market cullet
- No dilution of hot flue gas (to lower the temperature)
- No increase in particulates
- Minimize redox impact
- Minimize dry batch carryover
- Modular unit – requiring minimal modification to existing systems
- Solids/gas bypass mode to allow for uninterrupted glass production if Batch/Cullet Preheater requires maintenance
- Scalability to allow for system to increase in size for larger furnaces
- Minimize moving parts and maintenance requirements
- Minimize capital cost to offer attractive payback – target 18-36 months

This new batch/cullet preheater developed by Praxair is based on indirect heat transfer and can utilize the high grade heat from an oxy-fuel fired furnace without dilution air. The system concept is illustrated in the schematic diagram in Figure 4 below. The preheater can accept a mixed stream of batch/cullet at any ratio. The first stage of the preheater is an indirect radiative heat transfer section (RHRS) which the hot flue gas enters without dilution. Flue gas enters this unit at about 2500 °F and leaves at 1200 °F. If additional heat recovery is desired, an optional convective heat recovery section (CHRS) can be added to recover off-gas energy and reduce the flue temperature to 600 °F or lower. The preheated batch and cullet is then charged to the furnace.

In order to allow the plant to revert to normal operation in the event of maintenance to the batch/cullet preheater, the system will include a bypass mode to allow operation with the conventional batch/cullet feed system. During this period, the flue gas will be redirected to the existing stack to allow the preheater to cool down. By varying the damper positions, it will also be possible to operate the batch/cullet preheater in partial turndown conditions.

A 15 t/day pilot-scale unit (indirect, radiative section) has been built at the Praxair Technology Center in Buffalo and was commissioned in early 2007 and is shown in figure #5 below. Preliminary operating results from this system have been very promising. At design operating conditions, batch/cullet has been preheated to 900 °F-1000 °F and it is expected that this temperature can be increased with design optimization. At these conditions, flue gas exit temperatures on the order of 1200 °F have been measured, which also corresponds with design estimates. Field trials for the new batch/cullet preheater are expected to take place during 2008 on commercial glass furnaces at a 40-60 tpd scale.

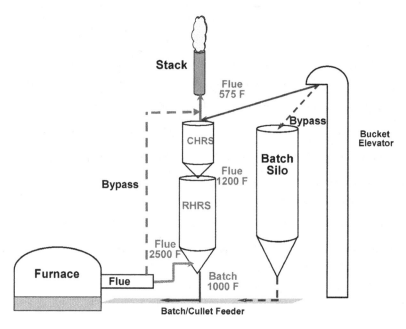

Figure 4: Schematic diagram of Advanced Batch/Cullet Preheater

Figure 5: Pilot Batch/Cullet Preheater at Praxair Technology Center

SUMMARY

About 30% fuel savings and significant reductions in CO_2 emissions, as compared with a state-of-the art regenerative furnace, have been demonstrated for an oxy-fuel fired container glass furnace equipped with batch/cullet preheating. Commercial experiences with two oxy-fuel fired container glass furnaces equipped with batch/cullet preheating demonstrated consistent specific fuel consumption of about 2.8-3.0 MMBTU/t at 570 °F preheat temperature with 50-60% cullet. The advanced batch/cullet preheater for oxy-fuel firing, which eliminates the air dilution for hot flue gas, has been demonstrated to preheat batch/cullet to 900-1000 °F and higher temperatures could be possible. With this new system, it is expected that the fuel consumption could be reduced to as low as 2.5 MMBtu/t, or by about 1 MMBTU/t relative to base oxy-fuel operation with a corresponding decrease in oxygen consumption. The savings from this system, combined with other environmental benefits anticipated, should dramatically improve the economics of oxy-fuel conversion.

REFERENCES

1. Kobayashi, H., "Advance in Oxy-Fuel Fired Glass Melting Technology", Proceedings of XX International Congress on Glass, Sep 26-Oct 1, 2004, Kyoto Japan
2. Kobayashi, H., K. T. Wu, G. B. Tuson, and F. Dumoulin, and J. Böllert, "Tall Crown glass Furnace Technology for Oxy-Fuel Firing", Proceedings of the 65th Conference on Glass Problems, The Ohio State University, October, 2004.
3. Kobayashi, H., Wu, K.T., Switzer, L.H., Martinez, S. and Giudici, R., "CO_2 Reduction From Glass Melting Furnace by Oxy-Fuel Firing Combined with Batch/Cullet Preheating", XX A.T.I.V. Conference - Parma (Italy) September 14-16, 2005G.
4. Lubitz, G., Beutin, E.F. and Leimkuehler, J., "Oxy-fuel fired furnace in combination with batch and cullet preheating", Proceedings of the NOVEM Energy Efficiency in Glass Industry Workshop, Amsterdam, May 18- 19, 2000.
5. Enninga, G., Dytrich, K., and Barklage-Hilgefort, H.J., "Practical Experience with Raw Material Preheating on Glass Melting Furnaces", Glastech. Ber., 65 [7] 186-191 (1992)
6. Schroeder, R.W., Snyder, S.J., Steigman, F.N., "Cullet Preheating and Particulate Filtering for Oxy-Fuel Glass Furnaces," Proceedings of the NOVEM Energy Efficiency in Glass Industry Workshop, Amsterdam, May 18-19, 2000.
7. Beerkens, R.G.C., "The Most Energy Efficiency Glass Furnace: Energy Efficiency Benchmarking and Energy Saving Measures for Industrial Glass Furnaces" , XXI A.T.I.V. Conference - Parma (Italy) September 20, 2005
8. Zippe, B.-H., "Reliable Batch and Cullet Preheater for Glass Furnaces", Glass technology, Vol. 35No.2, April 1994
9. Chamberlain, R.P., "Demonstration of a Batch/Cullet Preheater", Final Technical Report submitted to US Department of Energy, DOE-ID13386-4, June 21, 2001
10. Leone Industries: Experience with Cullet Filter/Preheater, 67th Conference on Glass Problems, The Ohio State University, October, 2006.
11. Schroeder, R.W., Kwamya, J.D., Leone, P., Barrickman, L., "Batch and Cullet Preheating and Emissions Control on Oxy-Fuel Furnaces," 60th Conference on Glass Problems, University of Illinois at Urbana, October 19 to 20, 1999.
12. U.S. Patent 5807418 (September 15, 1998)
13. Brouwer, F. and M.C. Sturm, "Gemengvoorverwarmer glassmeltoven 16" (translation – "Batch preheater glass melting furnace 16"), August 14, 1997 for NOVEM, The Netherlands proj. 391520/0011

INNOVATIONS IN CONTAINER GLASS PRODUCTION IN CENTRAL AND EASTERN
EUROPE

Günter Lubitz
Vetroconsult
Bülach, Switzerland

Looking at container glass production in middle Europe and North America means:

- glass packaging substitution by PET, cans and Tetrapack
- shrinking market segments for container glass
- over capacities
- furnace shutdowns
- plant closures
- shrinking sizes of container glass companies

But there is one company in Central Europe which has grown continuously. Within the last 20
years, it has developed from a mid-sized family-owned company in Switzerland to a leading
European container glass producer: the Vetropack group. To achieve this, a special
innovative approach was necessary. This paper describes Vetropack's original approach,
which has led to this successful development. The paper is divided into three parts:

I.   History and strategy of Vetropack Group
II.   Key factors for production improvements
III.   Successful plant modernization and integration

I. HISTORY AND STRATEGY OF VETROPACK GROUP

Founded in St-Prex at Lake Geneva in 1911, Vetropack is today number five in Europe and
market leader in Switzerland, Austria, Czech Republic, Slovakia, Croatia and Ukraine (see
Figure 1). Since 1985 sales have increased from 777 million to more than 4 billion pieces per
year. However, by the early 1980s Vetropack management was confronted with the following
obstacles:

- With Vetropack's market share of 85% the Swiss home market was saturated
- Full production capacities of the three Swiss plants at St-Prex, Bülach and Wauwil were
  only utilized for occasional exports
- There were disadvantages in production cost due to:
  - operation of three small plants
  - production lines which were dedicated to only small runs
  - high raw material, energy and labor costs in Switzerland
- It was realized that PET would partly replace glass as packaging material
- Vetropack was the leading container glass producer in Switzerland, but unknown in
  Europe and had no market connections
- Vetropack could only temporarily offset these disadvantages in production costs through
  higher productivity and higher quality, but the long-term potentials were exhausted

Based on these facts, Vetropack management realized that it had to find another strategy to be
successful in the long term. There were two possible directions for a new strategic orientation:

1. Introduction of new technology and products in an established market or

2. Expansion into new markets using existing products and existing technology

Vetropack decided on a double strategy:
- to invest in new PET bottle production technology for the Swiss market
- to expand with existing container glass production technology into a new market, like Austria in 1986 with the take over of Pöchlarn plant.

Vetropack Group implemented this double strategy from 1986 through 1996. The Group invested in an additional PET production capacity in Switzerland, and, additionally, expansion of the core activity – container glass productions – was continued in other counties.

In 1991 – after the opening of the eastern European border – Moravia Glass in the Czech Republic was acquired, and, in 1993, the second glass plant in Austria, the Kremsmünster plant. With the rapid development in the PET business and the two acquisitions, Vetropack had reached the limit of the investment possibilities of a family-owned company. Therefore, the PET business was sold in 1996, whose proceeds financed the expansion of the glass core business in Eastern Europe.

For the Swiss container glass industry, the sales-price-cost ratio became more and more unfavorable with the globalization of markets. The technical advantage in terms of productivity compared to the foreign competitors diminished further in the 1990s. Therefore, Vetropack was forced to adapt production capacities based on the market conditions in Switzerland. This meant that in 1993 the Wauwill plant, and in 2002 the Bülach plant, were shut down. Today in Switzerland only the St-Prex plant is in operation. However, expansion in eastern European countries was continued. In 1996, Vetropack took over the only glass plant, Hum na Sutli, in Croatia; in 2002 the Nemsova plant in Slovakia; and in 2006 the Gostmel Glass plant in Ukraine.

With these different takeovers, Vetropack gained entry into markets with exceptional growth. Therefore, Vetropack could establish in the European and eastern European markets a powerful position as a container glass supplier. The strategy of Vetropack can be summarized as follows:

- concentration only on container glass production
- maintaining the leading position in home markets
- organic growth in home markets through productivity improvements
- company growth through expansion
- concentration on regions with great economic growth in Eastern Europe

For the Vetropack group it is vital to bring quality, productivity, energy efficiency, environmental and safety aspects, and financial results to Vetropack level before the next acquisition step.

II. KEY FACTORS FOR PRODUCTION IMPROVEMENTS

Vetropack Group has grown through acquisition of existing glass plants. The acquired glass plants were on a poor technical level, production was ineffective, and therefore they were experiencing financial difficulties. Vetropack was forced to improve and invest in the new plants in order to integrate them rapidly into the group. According to our experience, the following key factors are important for production improvements:

a. Key figures and information systems
b. Good production practices
c. Good plant management and well-trained staff
d. Efficient furnaces
e. Properly-sized production equipment and floor plan

a. *Key figures and information systems*
Before starting optimization of a glass plant, evaluation of the existing performance must be made, so objective performance indicators must be implemented. Only then can performance be measured. Vetropack uses the following indicators to judge the plant performance:

- GCWT/BMD
- Efficiency
- Pack to Melt
- T1 and T2 Times
- Downtime
- Man-hours/GCWT
- Number of Customer Complaints
- Held Ware
- Customer Complaint Costs
- Specific Energy Consumption for Melting

Here are these indicators described in more detail:

Gross hundredweight (GCWT) per blow mold day (BMD) is an index for the IS machine speed. This index indicates how many gross hundredweight can be produced per blow mold day.

Efficiency in % describes the ratio between saleable pieces and cut gobs during net production time. Breakdowns of more than 30 minutes are neutralized.

Pack to melt in % describes the ratio between saleable tons and molten tons, impacted by a variety of stops when glass-flow continues.

T1 Time in hours is the time between last gob previous job and first gob new job (mechanical job change).

T2 in hours is the calculated production loss time between startup of new job and 16 hours after startup, at an efficiency of 95%.

Downtime in % is unplanned production interruption due to plant staff.

Man-hours/GCWT describes how many effective working hours (man-hours) are needed to produce one gross hundredweight (GCWT).

To evaluate plant performance it is also important to look at product quality. Therefore, certain figures are used, for example: Number of complaints is calculated from the number of complaints per 10 million pieces produced; Held ware is calculated in percent of the number of blocked wares from the total warehouse stock; Customers' complaint costs are calculated in per mill – which is the ratio between complaint costs and gross turnover.

The biggest cost factor in container glass production is no longer raw material cost, but energy cost. Therefore, a close follow-up of specific energy consumption, calculated in kcal per kg of molten glass for melting, is important.

b. *Good production practices*
After determination of the production performance with key performance indicators it is important to then introduce the second step in efficient production processes. This means:

- Documentation of all important production parameters
- Assurance of reproducible processes
- Definition of targets for all important parameters
- Definition of measures to reach the targets
- Close follow up of realization
- Preventive maintenance of production equipment

An example of good production practices is the job change process. Every minute of production interruption is a big loss due to the downtime of expensive production equipment. Therefore, job changes must be as short as possible. Special know-how is also required to start up production equipment and to reach, in the shortest possible time, a high productivity and quality level. Good organization and a sufficient number of well-trained specialists are a precondition for efficient job changes. Through job on and job off meetings it must be determined that all job change parts are available and well-prepared. In case of defined corrective actions from the previous job off meeting, it is important to follow up if they have been carried out and that the molds and job change parts were prepared accordingly.

Another example of good production practices is preventive maintenance. Due to 24-hour operation, there is permanent wear and tear of certain IS machine mechanisms and various IS parts. From past experience, and in combination with special checks prior to job changes, it is important to define exact exchange dates before productivity is affected in a negative way.

c. *Good plant management and well-trained staff*
In spite of modern production equipment, efficient container glass production still depends on qualified staff. The plant manager is a very key position and this person is required to have extensive knowledge in container glass production, as well as social competence combined with willingness to work hard.

Looking back to the development of the IS machine within the last 20 years, more and more pneumatically-driven parts have been replaced by servo-driven parts. Therefore, due to the complexity of operation and maintenance, demands on staff know-how have increased. In the past it was possible to work with semi-skilled workers. Today well-trained specialists are required. Therefore, fitters and electricians must be employed and trained to be machine operators or coldend operators. The same problem exists for section heads of mixing and melting, production, sorting, and quality control departments. It is recommended to employ young engineers from the areas of glass and ceramics, mechanics or electrics, and train them through internal training courses.

Furthermore, it is important to carry out continuous training. There are very few outside training courses available. Therefore it is important to carry out in-house courses on special topics with institutes like TNO or Glass Services.

Within a group like Vetropack with seven plants, good communication and exchange of information is vital. To ensure this, all executives of an acquired plant must learn the company language, which is English, quickly. There are excellent solutions for special problems in the different plants. These solutions must be identified and exchanged with other plants. Therefore special working groups, like furnace, production, and coldend, have been established, and delegates from each plant meet once or twice per year in different plants to discuss several topics. Another benefit from this rotation of meeting locations is that after a certain time all plants are well-known by all members of the working groups. According to our experiences, good education and continued training are important conditions for successful operation of a container glass plant.

d. *Efficient furnaces*
About 75% of utilized energy is needed for melting glass in a container glass plant. Due to the fact that energy prices have increased dramatically during recent years, it is important to build very energy- efficient furnaces. Another important aspect is to meet the requirements regarding more and more limited NOx emissions. 15 years ago Vetropack preferred the unit melter type furnace. Today mainly endport fired furnaces are used. Vetropack operates furnaces with a melting area from 44 m^2 to 156 m^2. Vetropack's approach is to build very energy-efficient furnaces and to regularly upgrade them to state-of-the-art technology. Therefore benchmarking studies were carried out together with TNO. In close cooperation with the German furnace construction companies, Horn and Sorg, it was possible to build more energy-efficient furnaces.

e. *Properly-sized production equipment and floor plan*
As already mentioned, Vetropack has grown through the takeover of existing glass plants. Every plant had its own technical history and philosophy. Therefore only in the case of furnace rebuilds or IS machine exchanges was it possible to introduce the Vetropack standard. Definition of Vetropack standard, planning and execution of investment projects is done by Vetroconsult in close cooperation with the different plants. Vetroconsult provides services in the fields of engineering, production quality and cost optimization for the Vetropack group. (These services are also provided for companies which are not in direct competition to Vetropack.)

Before planning a furnace with production equipment it is necessary to know the product range to be produced. Today a furnace should be
operated at least 10 years, and how is it possible to predict what will be produced in 10 years? This can be a difficult problem. But the solution is to design the equipment to be flexible. Also it is important that the production building is large enough for optimum efficiency and working conditions. If this is not the case, it is necessary to enlarge the production building.

III. Successful plant modernization and integration

Vetropack group took over the Nemsova plant in Slovakia in 2002 (see Figure 2). The Nemsova plant was a bankrupt company. The following will describe how the Nemsova plant has been developed and which measures have been carried out based upon the five above-mentioned key factors for production improvement.

Key figures and information systems
After introduction of Vetropack key performance indicators analysis showed that with the current performance, the cost structure and installed equipment at Nemsova could not survive (see Table 1).

At two furnaces 140 tons per day each were melted. The average energy consumption for melting was around 30% higher compared to Vetropack's best furnaces. 580 employees at five production lines produced around 91,000 salable tons per year. Tons packed were only 88%. Due to the low overall performances, production costs were accordingly higher.

Good production practices
In the next step, an audit of the production process was performed. The actual established process showed some differences compared to other Vetropack plants. Therefore, we adapted Nemsova's production process and introduced all relevant documents. We also ensured that all relevant production parameters were documented. In addition, we established action lists with clear targets. Reproducibility should be assured with all these measures. Also, preventive maintenance was introduced to reduce downtime. Also, based on Vetropack's experience, optimization of mold design and mold cooling was introduced to speed up production machines. It was possible to increase production speed from 2004 till 2007 in a significant way.

Good plant management and well-trained staff
Vetropack's philosophy is to give the local management full confidence and responsibility and the best possible support from our holding company, Vetroconsult and other plants. A close cooperation and support with Maravia Glass in the Czech Republic was established and a business partnership was created. Such an intensive integration of two companies consists of merging all relevant technical and commercial processes and intensive staff training. Within a short time it was possible to get an overview of the competence and motivation of key staff. Deficits in knowledge were compensated for through special in-house training courses, or training courses in other plants. Also, section heads of the different departments were integrated into the various working groups to benefit from group knowledge, or to bring in their own valuable knowledge. In addition, the plant organization was adapted to Vetopack's standard, and it was critical to put the most competent persons into key positions. In spite of adding one production line more, from 5 to 6, the number of employees has been reduced from 580 to 360 within 5 years. This is a 38% reduction. The whole integration process is very exhausting and needs the full support of all group members. Again, it is absolutely critical to take into consideration cultural realities, which include local laws and the mentality of the people. Furthermore, language problems can and will occur.

Efficient furnaces
Two furnaces were in operation at the Nemsova plant in 2002. The specific energy consumption for melting was 1,300 kcal per kg. Cullet ratio was 40%. One furnace was a cross-fired furnace and the other a recuperative furnace. The cross-fired furnace was at the end of its campaign and the recuperative furnace could be operated for another four years. Planning and installation of a new furnace was started immediately.

In the first big investment step, the 89 m^2 cross-fired furnace with a melting capacity of 190 tons per day was replaced by an 80 m^2 endport-fired furnace with a melting capacity of 230 tons per day. In the second investment step, a 64 m^2 recuperative furnace with a melting capacity of 140 tons per day was replaced by an 86 m^2 endport furnace with a melting capacity of 215 tons per day, without boosting. In both cases the metal line was raised; also the working ends were enlarged and the forehearths lengthened. As a result of both investments, between 2002 and 2007 specific energy consumption for melting was reduced by 21%, from 1,300 to 1,030 kcal per kg; cullet ratio was increased from 40 to 46%; and melted tons were increased from 100,000 to 150,000 tons. Furthermore, during this time NOx emissions were reduced from 680 to 160 tons per year by primary measures.

Properly-sized production equipment and floor plan
Analyses of plant layout showed that it was necessary to improve the arrangement to increase productivity. The two furnaces had been installed in two different buildings and it was decided to install the two new furnaces in one production building. An existing building was utilized and enlarged by 50 m, which resulted in a good layout with straight sorting lines.

In 2004, a new 10 section 5" IS machine was installed beside two overhauled and modernized existing smaller machines at the first new furnace. The coldend was equipped with three new sorting lines, some new inspection machines and a palletizer. An existing shrinking machine was transferred to the new furnace. Three production lines were installed at the new second furnace in 2006. Besides two overhauled and modernized IS machines, a new 10 section 5" machine was installed. The remaining equipment was transferred from the old furnace.

From 2003 to 2006, the total investment program had a value of €47 million. In the coming years it will be necessary to install additional warehouses and to replace the existing old recycling plant with a new one and to install an electro dust filter for both furnaces, due to legal regulations in 2009. As a result of this comprehensive investment program, production costs per ton of glass have been reduced by 30%. Table 4 shows the development of the Nemsova plant in important key figures from 2002 to 2007. 360 employees at six production lines produce around 136,000 good tons of container glass. This is equal to one good ton per employee per day, compared to 0.4 good ton per employee in 2002. This is an increase of 240%. Productivity counted in tons packed increased from 88 to 91%. Besides higher productivity, product quality has also improved. The Nemsova plant is able to fulfill the group requirement, called "One Brand One Quality".

FINAL REMARKS

Today it is still possible to earn money with container glass production and to grow as a company. The preconditions are:

- A good strategy
- A strong company philosophy
- Expert production know-how
- A willingness to invest continuously

Vetropack's approach is not short term cost reduction and dismissal of employees in an acquired plant. Our experience shows that production costs will decrease in the medium term if the focus remains on product quality, stability in the production process and qualified employees.

It must be remembered that political borders eventually lose their significance. True boundaries are not defined by national territories but by language, culture and history. Bridging these boundaries and discovering how to work together successfully can open up opportunities for companies to evolve and grow. Indeed, this remains the biggest challenge for management.

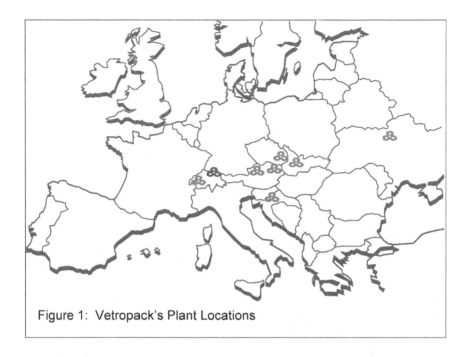

Figure 1: Vetropack's Plant Locations

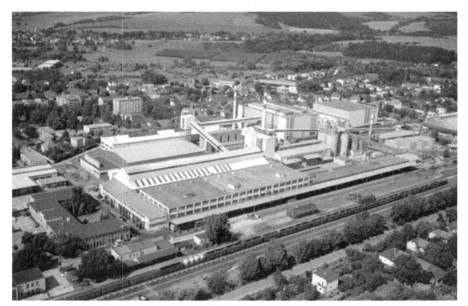

Figure 2: Nemsova Plant

Tabe 1: Status of Nemsova Plant in 2002

	2002
Number of furnaces	2
Ø Pull per furnace in tons per day	142
Ø Melted tons per day and plant	280
Ø Energy consumption for melting in kcal per kg	1,300
Ø Cullet ratio in %	40
Number of production lines	5
Good tons per day	246
Pack to melt in %	88.0
Number of employees	580
Good tons per employee per day	0.43
Dust emission in tons per year	31
SO$_2$ emission in tons per year	92
NOx emission in tons per year	680
CO$_2$ emission in tons per year	52,148

Table 2: Development of the Nemsova Plant from 2002 to 2007

	2002	2007	Deviation in %
Number of furnaces	2	2	0
Ø Pull per furnace in tons per day	142	206	45
Ø Melted tons per day and plant	280	411	47
Ø Energy consumption for melting in kcal per kg	1,300	1,030	- 21
Ø Cullet ratio in %	40	46	15
Number of production lines	5	6	20
Good tons per day	246	374	52
Pack to melt in %	88.0	91.0	3.0
Number of employees	580	360	- 38
Good tons per employee per day	0.43	1.04	242
Dust emission in tons per year	31	29	- 7
SO$_2$ emission in tons per year	92	36	- 61
NOx emission in tons per year	680	160	- 76
CO$_2$ emission in tons per year	52,148	59,500	14

TIN BATH BOTTOM BLOCKS – CHALLENGES AND NEW SOLUTIONS

Götz Heilemann, Bernhard Schmalenbach, Thomas Weichert
RHI GLAS GmbH
Wiesbaden, Germany

Klaus Santowski
RHI Refractories Technology Center
Leoben, Austria

INTRODUCTION

Currently, over 95% of the world's flat glass is produced using the float glass process introduced in the early 1960s, where molten glass is poured onto a bath of molten tin in a controlled atmosphere. This process replaced production by machine drawing or even blowing because it guaranties mirror quality glass without any grinding and polishing.

Due to the application of float glass in the building industry and automotive sector, the quality requirements for the glass are extremely high. Even the smallest glass defects distort the view and cannot be tolerated. All potential sources of glass defects have been carefully examined including the tin bath, which is one of the key sections in a float glass plant. As any corrosion of the refractories installed in the tin bath is a potential source of glass defects, this should be minimized.

The following paper describes the history of tin bath bottom blocks and the types of corrosion observed in the years since the start up of the float glass process [1]. Based on this history, a new ideal refractory guaranteeing minimum corrosion and interaction with the media such as the tin bath atmosphere, liquid tin, and molten glass was developed and is presented.

DEVELOPMENT OF REFRACTORIES FOR THE TIN BATH BOTTOM

In the tin bath bottom large refractory blocks are installed, termed blocking. The standard surface of the block is typically 960 x 605 mm², and the height varies between 305- and 152-mm. The standard block has four stud holes through which they are bolted down to the casing. A typical tin bath bottom block arrangement is illustrated in Figure 1. Since the beginning of the float glass process, these blocks have been based on fireclay; however, due to the different chemical and mineralogical compositions of the fireclay blocks, the problems detailed in the following sections have been observed.

Figure 1. Arrangement of tin bath bottom blocks.

Glassy Phase rich blocks with 25% to 30% Alumina (1950s)

Glassy phase rich fireclay with 25% (wt%) to 30% Alumina was installed for a short period; however, it was observed that Na_2O was transported from the glass through the molten tin to the surface of the fireclay block. Due to the low alumina and high silica content in the refractory material, feldspars (e.g., albite ($Na_2O*Al_2O_3*6\ SiO_2$)) and sodium silicates were formed. Some of these newly formed minerals melted below 800°C and as a result glassy phase drops formed on the blocking surface. Because of the low density of these silicate melt phases compared to molten tin these drops ascended to the tin bath surface, there these drops attached to or dissolved in the glass ribbon causing unacceptable glass defects. This phenomenon became known as the tadpole effect.

Blocks with approximately 40% Alumina (1960s)

Increasing the alumina content to approximately 40% eliminated the tadpole effect; however, after some years in service another problem was observed. Individual blocks within the bath cracked in the horizontal plane at the level of fixation (washer level) and a section of the block, with a thickness of about 180 mm, floated up. These floaters interrupted the glass production. Due to the thickness of the floating discs/plates, this phenomenon became known as the 7"-effect. The reason for this problem has not been resolved completely and the following theories are still under discussion:

- Too narrow expansion joints: The expansion joints were designed to prevent, as much as possible, tin from passing into the joints. However, the actual block expansion was found to be more significant than the theoretically calculated value and tension built up in the blocking, leading finally to cracking and detaching.
- Insufficient cleaning of the expansion joint during cold repair: In the 1960s the lifetime of the furnace was shorter than the lifetime of the tin bath, and the tin bath refractory lining was not changed for the second campaign. To avoid any pressure increase during the heat up, it was indispensable to clean the joints of tin. However, this was not 100% effective, resulting in increased stresses that led to cracking and floating.

- Cracks existing in the new blocks: During the 1960s the stud holes were drilled before the blocks were fired. During firing stresses increased in the washer level and cracking started. During the tin bath heat up and furthermore during the second heat up the cracks enlarged leading finally to complete detachment of the floaters.

It was assumed that these blocks were not flexible enough to compensate for the stresses and glass producers requested a material with higher flexibility. As blocks with approximately 40% Alumina contain about 30% glassy phase and as the glassy phase increases the brittleness of the blocks and decreases their flexibility, it was decided to develop a material with a lower proportion of glassy phase.

Blocks with 43% to 46% Alumina (1970s)

Around 1970, blocks with approximately 43% alumina and 15% glassy phase were introduced. The structure of the blocks is more flexible compared to fireclay with a higher amount of glassy phase. However, since this material is more sensitive to alkali corrosion another type of problem arose because material containing alumina can form the feldspar substitute nepheline when it contacts Na_2O rich phases. In addition, the amount of nepheline formed depends on the specific alumina content and reaches its maximum conversion into nepheline of 100% (complete conversion) with an alumina content of 45% (Figure 2). The formation of nepheline leads to the following consequences:

- A considerable volume increase from 20% with 60% nepheline to 35% with 100% nepheline (see Figure 2).
- A different thermal expansion of the fireclay and nepheline compared to the original fireclay

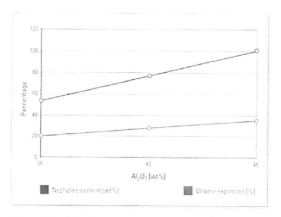

Figure 2. Nepheline formation and volume expansion as a function of the Al_2O_3 content.

By using material with approximately 45% alumina, block peeling became a problem after some years in operation. The detachment of nepheline shells started in the number 4 to 6 bays in direction of glass flow always after a change in settings of the tin bath for modifying the sheet glass thickness (i.e.; after a modification of the tin bath thermal profile). In the following months of operation, the peeling also moved to the number 2 to 8 bays, and during this time the surface of the peel enlarged considerably. In some cases the production had to be interrupted and an exchange of the entire bottom section appeared to be necessary. An example of a nepheline peel is illustrated in Figure 3.

Figure 3. Typical nepheline peel.

Blocks with 38% to 40% Alumina (1980s)

Since it was obvious that the quantity of nepheline formed in the glass blocks was dependent on the amount of alumina, the glass producers requested material with a lower alumina content. This type of material with an alumina content of 38% to 40% has been in service from the beginning of the 1980s. Whilst in some cases campaigns of 12 years or more have been reached without problems, in most cases nepheline peeling has occurred. However, the time period between installation and first peeling has varied between 5 and 10 years.

Nepheline Peeling

As nepheline peeling is still the most common problem encountered with tin bath bottom refractories, the corrosion mechanism was investigated more comprehensively. The following three different block types were analysed after use and the physical properties and chemical compositions are detailed in Table 1:
- Block 1: Initially contained 40% alumina and approximately 30% glassy phase, showed no peeling.
- Block 2: Initially contained 43% alumina and 15% glassy phase, showed peeling. This block had a rather high gas permeability of 8nPm.
- Block 3: Initially contained 46% alumina and 15% glassy phase, showed peeling. This block had a rather low gas permeability of 2,5nPm.

No.	Sample location	D (g/cm³)	Po (%)	GP (nPm)	Al₂O₃ (wt%)	K₂O (wt%)	Na₂O (wt%)	Glassy phase (wt%)
1	Block (no peeling)							
	0–10 mm	2.29	15.1	-	35.8	0.9	10.6	
	280–300 mm	2.11	21.5	1.9	40.3	1.0	0.1	30
2	Peel (peeling)							
	0–8 mm	2.28	16.2	-	37.1	0.4	11.2	-
	Calculated	-	-	-	42.6	0.5	0.1	-
	Typical	2.18	19	~8.0	43.0	0.7	0.1	15
3	Peel (peeling)							
	0–6 mm	n.t.	n.t.	-	39.9	0.3	9.7	-
	Calculated	-	-	-	46.2	0.3	0.1	-
	Typical	-	-	~2.5	-	-	-	15

Table I. Postmortem analyses of fireclay tin bath bottom blocks including the physical properties and chemical composition of two samples (infiltrated material 0–10 mm and original material 280–300 mm) taken from a nonpeeled block (1) and the peel from peeled blocks (0–8 mm and 0–6 mm, samples 2 and 3 respectively). The "Calculated" values are without infiltrated oxide and "Typical" indicates the standard block composition. Abbreviations include bulk density (D), open porosity (Po), gas permeability (GP), and not tested (n.t.).

Based on these results, the following conclusions were drawn:

- The gas permeability has no influence on the alkali absorption. This fact contradicts a theory of some glass producers that a low permeability prevents deep alkali infiltration. According to this theory only thin layers would not been able to detach from the block.
- All blocks absorb a similar quantity of alkalis, and nepheline is formed in all blocks. The typical alkali infiltration and nepheline formation at the block surface is illustrated in Figure 4. However, it is only the blocks with a low amount of glassy phase that show peeling.

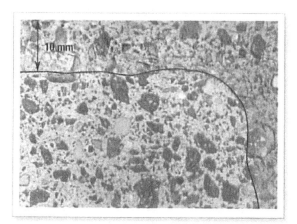

Figure 4. *Typical infiltration and nepheline formation at the block surface.*

Analyses indicated that the glass phase plays an important role in the peeling of the fireclay refractories and that it is able to act as a gluing agent in the block. The nepheline layer that is always formed is attached to the brick and although the thermal expansion is different, the layer is not able to loose contact with brick. It is also apparent that a minimum level of approximately 10% to 12% glassy phase is necessary to bind the nepheline layer. Finally, it is established that during service the glassy phase crystallizes and is thereby its volume is reduced. The degree of reduction is dependent on the working conditions of the bath, in particular the temperature and the temperature variations, and my vary from 0,5% to 2 % per year.

The campaign length until the first peeling occurs is therefore dependent on the original amount of glassy phase in the blocks and the degree of its reduction. Some tin bath bottom blocks start to peel after very short campaigns (i.e., 4 years). These blocks normally contain only a small amount of glassy phase (i.e., 15%) and are operating at high temperatures with many temperatures variations. Other blocks can reach campaigns of 15 years or more if they have a high glassy phase (i.e., 30%) and the tin bath is operated without any temperature variations with coloured glass. Therefore, if long campaigns need to be achieved without peeling, a material should be installed with the maximum amount of glassy phase. This explains why the development of a material with a lower portion of glassy phase failed in operation although.

Bubble Formation

In some tin baths, bubble formation (bubbling) has been observed to a large extent. As the bubbles destroy the quality of the glass ribbon their formation must be avoided. Typically, the reason for the bubble formation is due to thermal transpiration of the blocks. If the mean diameter of the block pores is smaller than the diameter of the gas molecules in the atmosphere, thermal transpiration starts. The gas is sucked from the lower, colder side of the blocks to the hot side and the thereby formed bubbles float up to the glass ribbon and form dents.

To characterize the potential for thermal transpiration the value of hydrogen diffusivity was introduced. It describes the effect of migration and the pressure build up of gases from the cold outside towards the hot inside, and a corresponding test has been introduced. If in this test, values of < 150 mm water gauge (WG) are reached, the block is considered to have no potential for thermal transpiration. However, it must be taken into consideration that a block is not homogeneous and that some parts may have higher values, even if the sample tested is under the critical value. For this reason, values of < 100 mm water gauge are not considered critical.

The relationship between permeability, hydrogen diffusivity, and pore size are depicted in Figure 5, which illustrates that to attain a low permeability the pore diameter must be decreased. In extreme cases (i.e., with a permeability of < 1 nPm), the diffusivity in terms of water gauge (build-up of pressure) reaches values of 120 mm WG or more, resulting in block bubbling.

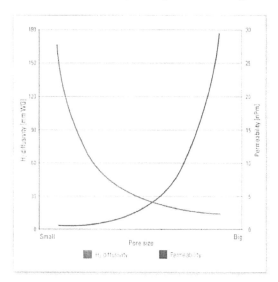

Figure 5. The relationships between hydrogen diffusivity, permeability, and pore size. No transpiration occurs with a pore size < 1 μm.

The possibility of bubbling formation increases with an increased temperature difference between the hot and cold side of a tin bath bottom block due to higher gas movement. If special glass is produced, requiring a higher temperature level in the tin bath, the gas pressure can lead to bubble floating at lower hydrogen diffusivities than observed for normal float glass (soda-lime-silica) processes.

Conclusion for Fireclay Tin Bath Bottom Blocks

Based on the results discussed, an ideal fireclay tin bath bottom block has the following characteristics:

- Alumina content of approximately 40%.
- Glassy phase content of approximately 30%.
- Cold crushing strength of approximately 50 N/mm² to ensure a good surface appearance with low risk of damage during transport and installation.

- Hydrogen diffusivity in the range of 50 mm WG.

Additional strength properties such as the modulus of rupture and the flexibility of a material indicated by the Young's modulus are currently considered less important, because the reasons for crack formation and breakages have been investigated and are not influenced by the material cold properties.

The alkali absorption curve for a fireclay tin bath block exposed to tin containing dissolved alkalis, in a reducing atmosphere, is measured by scanning electron microscopy (SEM) and energy dispersive X-ray (EDX) microanalysis, and is illustrated in Figure 6 [2]. The analysis took place from the surface down to 5 mm deep into the material. It demonstrated clearly that the fireclay block had absorbed Na_2O and a reaction had taken place between Na_2O and the block. It also illustrated that the bonding matrix principally absorbed the alkalis whilst the coarse grains were affected to a much lesser extent. Therefore, it appears that a block with the glassy phase concentrated in the bonding phase resists alkali attack better than a block with an equally distributed glassy phase or when the glassy phase is mainly concentrated in the grains. This means that the glassy phase forming agents in the feldspar such as K_2O and Na_2O have to be concentrated in the fine part of the block forming the bonding phase after firing.

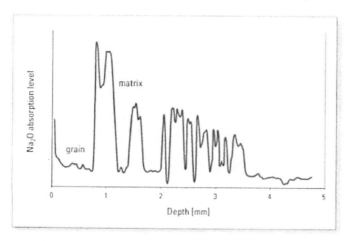

Figure 6. Na_2O absorption curve of fireclay blocks from molten tin. The analysis was performed using SEM-EDX.

SUPRAL 40FG is a fireclay block with the aforementioned characteristics. The chemical composition and most important physical properties of this type of material are detailed in Table 1 (No.1).

Blocks based on Calcium Aluminate
As blocks based on fireclay form nepheline during operation, it was clear that a completely new material had to be developed to address this problem. Investigations with hydraulic bonded material revealed that the calcium aluminate bonding phase was not attacked by alkali compounds. This led to the development of the novel RHI Refractories grade SUPRAL CA, consisting of nearly 100% calcium aluminate, which fulfils the necessary requirements regarding the chemistry and physical characteristics for tin bath bottom lining materials, including:

- No chemical reaction with tin: The result of a test where molten tin was filled into a crucible of the new material is depicted in Figure 7. After 100 hours at 1000°C under reducing conditions, no infiltration of tin into the crucible was visible, and no reaction had taken place.

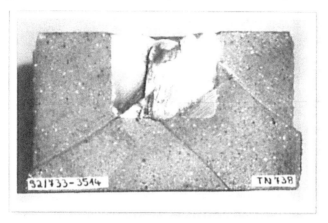

Figure 7. *Molten tin penetration test with SUPRAL CA performed at 1000 °C for 100 hours under reducing conditions.*

- No reaction with alkali components dissolved in the tin bath: A comparison of the previously described Na_2O absorption curve of fireclay blocks with the Na_2O concentration profile on this new grade is provided in Figure 8. It shows clearly that the absorption and reaction described for the fireclay block does not occur in the case of the new material. There is practically no absorption by this block and consequently only very minor reactivity is observed . The results of the alkali corrosion test of two SUPRAL grades are illustrated in Figure 9. Specimens of SUPRAL CA and SUPRAL 40FG were placed on a platinum crucible containing sodium as the reactant. The crucibles were heated to 950°C and after 96 hours the difference was clearly evident. Whereas the fireclay material showed strong nepheline formation, the new grade was only infiltrated by the vapour without any reaction occurring.

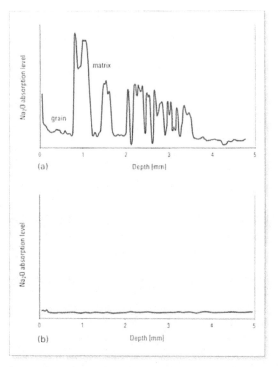

Figure 8. A comparison of the Na_2O absorption curve from molten tin by (a) fireclay blocks and (b) SUPRAL CA. The analyses were performed using SEM-EDX.

Other observations:
- No influence of the reducing atmosphere present in the tin bath.
- Good mechanical strength to enable perfect grinding and drilling as well as to eliminate handling damages: A smooth surface and sharp edges are obtained after grinding and drilling.
- Thermal expansion comparable to fireclay blocks: The design of the expansion joints does not have to be modified when replacing the fireclay with the new material.
- Thermal conductivity is lower compared to fireclay material: The thermal cooling can be reduced to reach the same casing temperature and a similar thermal gradient or reducing the block thickness can also be considered to obtain the same result.
- Thermal resistance is higher in comparison to fireclay blocks: Therefore the new material can work at a maximum temperature of 1200°C, whereas fireclay starts to creep at 1100°C.
- Hydrogen diffusivity is at a very low level of approximately 10 mm WG: The formation of bubbles under the glass ribbon by thermal transpiration during the campaign can be excluded completely.

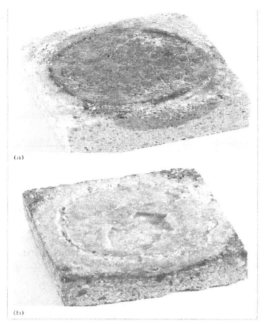

Figure 9. Alkali vapour test of (a) SUPRAL CA and (b) SUPRAL 40FG performed using sodium heated in a crucible at 950 °C for 96 hours

The chemical composition as well as the physical characteristics of SUPRAL CA are detailed in Table II.

Chemical composition (wt%)

SiO$_2$	5.0
Al$_2$O$_3$	68.0
CaO	24.0
MgO	1.5

Physical characteristics

Bulk density	2.26 g/cm³
Open porosity	23 vol%
Cold crushing strength	55 N/mm²
Refractoriness under load t$_{0.5}$	1350 °C
Young's modulus	12000 N/mm²
Modulus of rupture	9 N/mm²
H$_2$ diffusivity	10 mm WG
Gas permeability	7 nPm
Thermal expansion at 1000 °C	0.65%
Thermal conductivity at 1000 °C	1.0 W/Km

Table II. Chemical composition and physical characteristics of SUPRAL CA.

The new grade consists of various calcium aluminate phases with different hydration affinities; therefore, it was expected that during cutting, grinding, drilling with water the blocks would re-hydrate. The chemical bonded water could dissolve during operation and provide a source of bubble formation. However, investigations performed on finished blocks generated the following results:

- Fired SUPRAL CA is able to absorb a maximum of 5% water if placed under vacuum.
- In practice approximately 3% of the moisture is bonded as re-hydrated crystal water.
- During heating up the moisture is pressed to the cold side of the block.
- The moisture disappears at 350°C (i.e., the hot side of the block is free of water).
- The moisture is enriched at a defined level and location near the cold side of the block and is bound as crystal water that is not able to migrate to the hot surface.

The results were confirmed when the first blocks of the new material went into service. There was no bubble formation in the first year(s) of the campaigns and bubble formation is not anticipated during the ongoing campaign.

It should be highlighted that the new material is not only a reliable solution for long-term tin bath bottom linings, but also for the bottoms of tin baths where special glasses are produced at high temperatures (such as borosilicate and aluminate glasses for displays).

SUMMARY

This paper presents a summary of the history of tin bath bottom blocks and describes the reasons for the most common glass defects caused by these blocks, which include:

- Nepheline peeling
- Bubble formation

This paper discusses the development and composition of an optimized fireclay block that is able to operate in a tin bath for campaigns of up to 10 years without causing problems. However, as current campaigns of 12 to 15 years are typical in the float glass producing units, a material must be installed that is able to guarantee trouble-free behaviour over this long time period. Therefore, a new material was developed based on calcium aluminate: the grade SUPRAL CA. Laboratory tests confirmed its superior resistance to the bath conditions, and the first installations in the complete hot end of tin baths were performed: the initial heat up occurred without any problems.

Furthermore, during the first couple of years of operation no particular problems were observed. Based on these positive experiences, float glass producers worldwide have ordered the blocking for several further projects.

REFERENCES

[1] Wieland, K., Weichert, T. and Routschka, G. Zinnbadbodensteine. Presented at the XXXVI International Colloquium on Refractories, Aachen, Germany, 27-28 Oct. 1993; pp. 101-105.
[2] Wieland, K. and Weichert, T. Refractories for the tin bath in float glass plants. Veitsch-Radex Rundschau. 2000, 2. pp 29-40.

ELECTRIC FOREHEARTHS FOR BOROSILICATE GLASSES – A RECENT VIEW

Peter R.H. Davies and Alex J.R. Davies
KTGSI
Wexford, Pennsylvania

Douglas H. Davis, Christopher J. Hoyle, and Larry McCloskey
Toledo Engineering
Toledo, Ohio

I. ABSTRACT

The forehearth or front-end system performs an important function filling the gap between the batching and melting steps and forming. This function consists of 1) simple delivery, 2) extraction of excess heat added for melting and refining, 3) reaching thermal uniformity, 4) and being at the correct temperature. Joule-heated electric forehearths accomplish these well.

However, the capital cost of an all-electric forehearth can be 1.5 - 4 times that of a gas-air system. Nevertheless, there are operational benefits that can offset this disadvantage.

Due to very efficient energy transfer into the glass, joule-heated forehearths can provide lower operating costs per ton of glass than with natural gas, depending on local energy pricing. The sensitivity of operating cost to energy pricing is examined.

In addition, joule-heated forehearths essentially eliminate emissions. There has been growing concern over uncontrolled emissions into the plant environment and outside, both from combustion gases and volatiles from the glass. We will discuss technical difficulties and potential benefits of all-electric forehearths for E-glass operations.

The use of computer CFD-modeling and salt bath studies are used to ensure appropriate design, as is full-load testing of all electrical systems.

II. INTRODUCTION

The combination of channels, heating system, and control devices that cool, condition, and deliver the glass from the exit of the melter to the forming bushing has various names depending on the owner and the builder, e.g. forehearth, front-end system, a combination of the two, etc. For simplicity, we will use "forehearth" to describe the complete unit, although you will note that this term is used sometimes for the immediate section feeding the bushings.

The forehearth must deliver X tons of melted glass from the furnace to the forming devices. In addition, however, the system needs to cool this glass to the desired forming temperature with close tolerances in both temperature and chemistry. In the fiberglass sector, some are using the joule-heated electric forehearths to accomplish these goals.

The capital cost of the electric forehearth is greater than that of a gas-fired system due to electrical power supplies and equipment. However, more efficient energy transfer into the glass can give lower operating cost where there is favorable local energy pricing. In addition, essentially zero emissions from the forehearth can be an important benefit to the fiberglass plant.

We will provide an overview of the design, operation, and benefits of all-electric forehearths for C-Glass, as well as a brief discussion for E-glass. We will show how both CFD (Computational Fluid Dynamics) and physical modeling help us scale tonnage and avoid operational problems.

III. BASIC PRINCIPLES OF ELECTRICAL GLASS HEATING

When an electrical potential is applied across molten glass, the most loosely bound components in the glass (normally alkali ions) move through the glass under the influence of this potential. This "current" passing through the glass generates heat. As in most conductors, the heat generated is described by Joule's first law, and in turn using Ohm's Law as,

$$\text{Heat Generated} = I^2 R\, t = V I t$$

where I = current, V = applied voltage, t = time, and R = resistance of the material.

Glass differs from most conductors in that its conductivity is highly temperature-dependent. At room temperature, the resistivity is for all practical purposes infinite; the glass does not conduct electricity, and therefore no heat is generated. As the glass is heated (by some other means), the alkalis are more able to move through the structure, giving a "current", and this generates some additional heat internally.

Therefore, when electrodes are inserted into molten glass and a voltage is applied between the electrodes, current will flow through the glass. The resistance to this current will create internal heat within the glass; normally described as Joule-heating. Temperatures immediately adjacent to the electrodes will normally be higher than in the bulk glass, but the electrodes themselves are not creating heat. The electrode is just a means for applying the potential drop (voltage) across the glass.

IV. FOREHEARTH GENERAL DESCRIPTION

The size, capacity, and length of a forehearth front-end system can vary greatly. In the fiberglass insulation industry, the forming end of the plant is where a series of bushings feed glass to spinners over the collection belts. There may be either one or two bushing sections, and, depending on the plant geometry, the length from the melter to where the bushings need to be located from as little as 60 feet up to 250 feet. However, the requirements are the same.

A forehearth system will typically be broken down into three functional sections, i.e. cooling zone, conditioning zone, and the maintenance zone. Names for the various sections vary by company, but the functions are the same. As shown in Figures 1 & 2, channel width in the various sections becomes increasingly wide from the bushings back to the melter, so that in spite of viscous resistance to flow, there will be an adequate head of glass over the bushings.

Figure 1
Variation of Glass Depth and Width along Forehearth

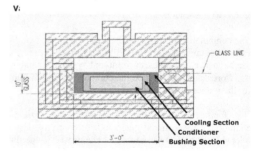

The refractory section in Figure 1 is in the cooling section (or alcove) where the glass cross-section is the greatest.

As discussed later, there is a wide variation in C-glass compositions, and the temperature to provide the log 3 viscosity normally desired for forming could vary from 2000^0 F–2100^0 F (1093^0 C-1149^0 C). For discussion, we will use a generic glass composition and viscosity.

Figure 2
Forehearth Front End System –C -Glass

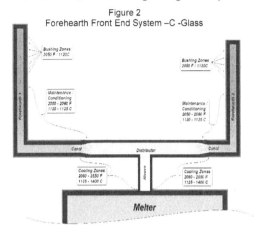

The first section (Cooling Zone) is where the excess heat from the high temperature needed to melt the batch and refine the glass can be dissipated through the refractory. This is seen in Figure 2 below, where a typical, but arbitrary, example of a larger, dual forming-zone forehearth system is shown. A typical heat loss in this area is from 1400^0 C down to 1125^0 C (275^0 C drop). Little or no power is applied to this zone during normal operating conditions. However, they are equipped with the largest of the forehearth transformers. This is because, in the event of a hot hold situation, with no incoming hot glass, high power inputs will be needed to overcome the designed high heat losses and maintain temperature.

The second section would be the conditioning zone. With proper design and operation, the glass arriving here will have an average temperature in the 1125-1120^0 C range, although due to the need for rapid dumping of heat in the prior zone, the specific temperatures will be far from homogeneous. In this zone, time will be provided, and compensating heat where needed, to equilibrate the glass to within 5 C of the desired forming temperature.

The third zone is that just before and above the bushings delivering glass to the fiberizing spinners. The desired forming temperature for a typical C-glass would be 1120^0 C, and the thermocouples above each bushing should be within 1° C of this temperature.

Figure 3
Short Forehearth with Single Forming Zone

Each zone will consist of at least one electrode circuit. Conditioning and bushing zones are similar in construction. The refractory construction of these zones is designed to minimize heat losses while maintaining glass temperature. Under ideal conditions, very low power inputs will be required to maintain the desired

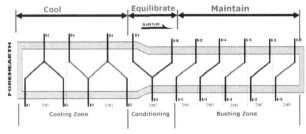

temperature. Bushing zones will typically be the smallest transformers in the forehearth system. Each bushing (drain to spinner) will have a dedicated electrode circuit (normally two electrodes), and each dedicated circuit will be controlled as an individual control zone. A conditioning zone, however, may contain multiple electrodes as part of a control circuit.

Figure 4
Typical Joule-heated Forehearth

The overall forehearth can consist of as few as three individual electrical circuits, but may be scaled to meet customer's needs. Figure 3 shows a relatively short forehearth system that supplies a single forming zone with five bushings. Figure 4 shows a view of a typical section of an all electric forehearth.

Every electrical zone in the forehearth system has a control thermocouple. Each of these electrical zones is designed to maintain the glass at a given temperature set point. While there are several ways for the power input to be controlled (see below), the control computer will adjust the power being applied to the zone to maintain the temperature set point.

V. FOREHEARTH POWER CONTROLS

All forehearth systems are designed to be operated as stand-alone systems so that they can be independent from maintenance or diagnostics on the larger system. This is why most forehearth designs incorporate some level of PLC controls. Even in the event of a computer failure, the individual zones of the forehearth can be operated locally by placing the control switch to LOCAL and adjusting the SCR output manually.

The SCR cabinets contain primary voltage isolation breakers, SCR's and various other devices. Indication of the zone electrode currents and voltages is located on the front of each panel. Controls can be digital and/ or analog depending upon customer comfort.

The energy dissipated into the glass (power) may be controlled in five ways. They are temperature, current, resistance, voltage, and power control. Temperature control is the preferred method of control in a forehearth during normal operation. Based on the differential from temperature set point, voltage is adjusted appropriately, although current limit settings prevent excess amperage. Current control of power input is very useful during unstable conditions (for example during startup) as it will decrease heat input to the glass as temperature increases and increase heat input as temperature reduces.

Resistance control is essentially the same as temperature control as resistance varies with the glass temperature. This is not a normal method of control as it requires some calculation and as it is better suited for longer, single circuit firing path. Voltage and power control are not normal methods of control but can be used in special conditions.

Figure 5
Top Change on Cabinet

VI. TRANSFORMER TAP SWITCHES

Each zone also contains a "tap position change alarm" that will indicate that the transformer is operating inefficiently. Maintaining the appropriate applied voltage as close as possible to the maximum transformer voltage setting (maximum Power Factor) is an important tool for keeping the forehearth at its minimum cost per ton of glass. These systems both remind the operator when a tap change would be appropriate, and make the transformer tap change simple and fast.

Figure 6
Transformer Tap Change Diagram

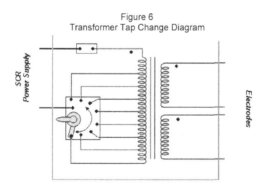

Each transformer is equipped with at least one off-load tap switch as shown in figure 5 and 6. These tap switches allow the operator to select the transformer output voltage to maintain as high a power factor as possible. Correct operation of the tap switch will significantly increase the efficiency of the plant. A tap switch change light (alarm) is provided to help operators maintain efficient operations of the transformers. While these systems are intended to encourage tap changes for efficiency improvement, inappropriate tap changes could cause electrode currents to exceed the transformer design specification and reduce its operating life.

VII.   ELECTRODE CHOICE AND PLACEMENT

A.   Molybdenum is the Choice

The designer has various options when assembling the heating systems for the electric joule-heated forehearth. However, with respect to properties and cost for an electrode material, the only practical choice at present is molybdenum, especially for the borosilicate glasses. It is extremely rigid at high temperatures, although easily oxidized. These high temperature oxides are volatile, but the slightly reduced state of the glasses of concern here will be adequate to prevent oxidation by the glass during operation. Steps are taken to avoid oxidation before glass fills the forehearth, and care with obtaining a glass seal in the refractory face will minimize oxidation of the electrode tail in the wall.

B.   "Dry" Versus "Wet" Electrodes

However, one does have to choose between "dry" and "wet" electrodes. The water-cooled electrode holder can be installed in the sidewall or bottom of the forehearth. The water-cooled sleeve provides a positive seal of frozen glass, as a guard against glass leakage and oxidation of the molybdenum rod in the wall that could cause future failure. At high temperatures and with aggressive glasses, the electrodes would experience some wear over the years; losing metal to the glass by oxidation and

Figure 7
Water-Cooled Electrode Holder

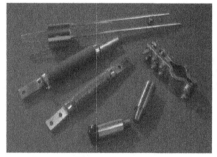

Figure 8
Electrode Connections

diffusion, and the water-cooled sleeve facilitates periodic advancement of the electrode into the glass to maintain the original path length, and maintaining the volts-to-amps relationship. The choice depends on the expected front-end system campaign life. If this is expected to be long, then electrode advancement might be needed and the water-cooled electrodes will be used in the entire system. Figure 7 shows a typical water-cooled holder. There also can be a connection for an inert gas flush within the holder, which is seldom used in forehearth applications. Figure 8 shows some of the various electrode connections.

The electrodes can be started up flush with the end of the water-cooled holder to avoid oxidation during heat-up. The electrodes will be pushed in and powered up when glass level is up, covering the electrodes.

Figure 9
Dry-Type Electrodes

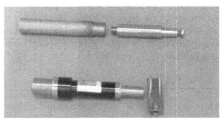

Figure 9 shows a dry-type electrode, which can be inserted directly through the forehearth, walls, with no intermediate device. They are used where further advancing is not expected. Dry-type electrodes present a lower heat load on the total system, only losing energy by conduction through the rods to the outside. In case of failure, however, replacement of these dry electrodes is difficult, often requiring hot drilling-out of the electrode.

Dry-type molybdenum electrodes can also be inserted flush with the hot face and inserted later. However, with long forehearth systems the labor and time to do so are significant. Therefore in most cases the electrodes are coated with a temporary oxidation-resistant material, inserted directly through the refractory walls, and extended into the forehearth channel from the start. Without the oxidations protection, these electrodes would be largely consumed during heat-up before they are covered by glass and protected from further oxidation.

C. Electrode Placement

The all-electric forehearth design must be carefully designed to provide a uniform heat input, accounting for the pull rate and the required temperature. In order to maintain electrode life, electrodes are spaced and powered to keep amperage loadings below a certain design maximum amps/square inch of electrode surface under normal working conditions. Higher amperages can lead to overly high electrode surface temperatures, which increases the diffusion losses of molybdenum into the glass. Experience to date shows that with careful design, careful attention to surface loading, and attentive supervision there is no need to advance electrodes to accommodate wear for at least seven years of operation.

Planning is underway to offer electric joule-heated distribution channels for use on E-glass furnaces. These glasses experience significant boron losses, second only to the C-type glasses. However, the electrical resistivity of the E-glasses is very high, having alkali contents of generally 1% or less. This requires higher operating voltages and less separation between electrodes (more electrodes) to generate the heat-producing electrical currents in the molten glass. As we will see below, special care with the refractories will be required to minimize current flow through the refractories. Physical modeling of these systems continues to be important to verify that the proper electrical relationship is maintained. Numeric computer-based modeling has been increasingly used to optimize the design. We anticipate that these electric joule-heating forehearth systems will be very useful to the E-glass industry.

VIII. REFRACTORY CHOICE – RESISTIVITY IS IMPORTANT

As in most engineering exercises, the design of a forehearth involves a number of educated compromises. As always, we have to decide between the cost of the forehearth and the campaign life. However, in the electric forehearth we also have to factor in the electrical resistivity of the refractory itself. Some of the best refractories for resisting the corrosion by these borosilicate glasses are also quite electrically conductive. Secondary currents through the refractories instead of the glass are not effective in heating the glass and can seriously reduce the life expectancy of the refractories and reduce glass quality.

A. High-Alkali Borosilicate – C-Glass

C-glass is the common designation for a high-alkali borosilicate glass. The composition shown in Figure 10 is roughly typical of a C-glass. The variations in commercial compositions of C-glasses are surprisingly large.

In Figure 11, the electrical resistivity of various glasses of interest and some of the refractories used are shown. At the bottom of the figure (low resistivity), is shown the resistivity of two typical C-glasses of the type used to produce insulation wool.

In the construction of forehearths for these C-glasses, we frequently use materials from the SERV family, pressed materials

Figure 10
High-Alkali Borosilicate C-Glass
SiO_2 - **62.75 %**
Al_2O_3 - **3.0**
B_2O_3 - **6.75**
CaO - **7.5**
MgO - **3.25**
Na_2O - **15.75**
K_2O - **1.00**

with high chrome contents that have shown reasonable life for the cost. The SERV 30 DC (30% chromic oxide) is usable in some low corrosion areas, and offers no competition to the glass as an electrical path. More commonly used in forehearths (for their improved corrosion resistance and longer life), however, are the SERV50DCX and the SERV52XL materials. Significantly more corrosion-resistant yet, but also significantly more expensive, is the SERV95 material. At the forehearth operating temperatures of roughly 1300°C, the last three SERV materials have roughly the same relative resistivity; much closer to that of the glasses. However, 1) since these refractories are still 5-10 times more resistive than the C-glasses in question, and 2) since the electrical path between the electrodes in the glass being much shorter than the alternative paths through the refractory, and 3) since the refractory is somewhat cooler than the glass; the potential for an electrical "leakage" through the refractory is not a big concern for melting C-glass.

Fiaure 11

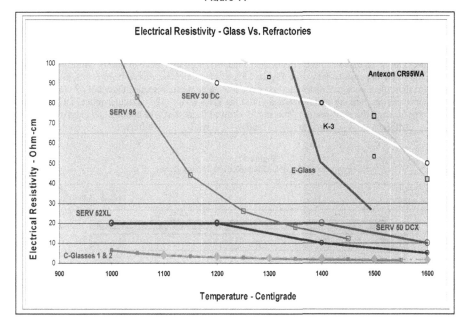

B. E-Glasses

However, looking at the resistivity curve for a typical E-glass also shown in Figure 11 above, we have quite a different picture. The materials we discussed as suitable for the C-glasses could now present quite a viable alternative pathway for current. Several other materials, including fused-cast K-3 (the standard for many DOE hazardous-waste melters) and Antexon CR95WA (a 95% chromic oxide isostatically-pressed material) and similar materials offer reasonable corrosion resistance and a safer electrical design mode.

SEPR zircon-bonded refractories have been useful in E-glass forehearth operation as electrode blocks and bottom paving blocks with higher resistivity than the dense chromes. ZS-1300, ZS-78, and ZS-835 have been useful here.

IX.   USING MODELING FOR ALL-ELECTRIC FOREHEARTHS

A.   Physical Modeling

Physical modeling of these systems using conductive liquids continues to be important for electric forehearths. With the relatively small dimensions of the forehearth systems, exact positions of the electrodes are critical to maintaining the ultimate amperage and voltages within practical ranges.

Transformer specifications are dictated by the results of the physical modeling. We perform this physical modeling for every project.

B.   CFD Modeling (Computational Fluid Dynamics)

We increasingly use numeric computer-based modeling to optimize these designs. The ease with which geometries can be changed and the effects calculated, offers an assurance of highly uniform, delivered glass. The modeling results shown on Figure 12 are for an alcove, which receives hot glass from a riser, and then must cool and distribute that glass into separate forehearths. One-half of the system is shown here and are looking at a horizontal cut at the glass level. The computer model includes both the combustion and the glass portion. This model allowed us to "test" the cooling and glass distribution at   different   tonnages   and   with   different   configurations.

Figure 12
Computer Model of C-Glass Alcove (Cooling Zone)

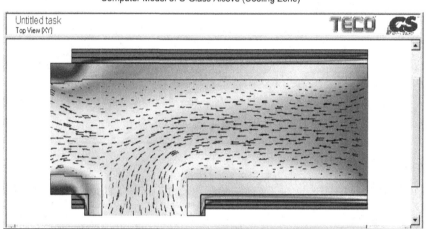

X.   ENVIRONMENTAL CONSIDERATIONS

A. Venting the Forehearths into the Plant is becoming a Problem

There is growing concern over the common practice of venting the forehearth combustion gases and volatiles from the glass within the manufacturing building. The lack of emissions from an all-electric, joule-heated forehearth can be of significant importance to the plant operation in this regard.

Compared to losses from the melters, the industry might be excused for having ignored those of the forehearths. The venting of gas-fired forehearths (using cold combustion air) into the plant area will have much lower NOx levels than the waste gases from the melter itself. In addition, the volatile / evaporative losses from the relatively cool glass in forehearths are not nearly at the level we see in gas-fired melting furnaces.

However, authorities continue to question the effect of waste gases vented into the plant working area from the forehearths/front end systems. Fiberglass forehearths/front-ends can consume a surprisingly large amount of energy due to the sometimes-huge length and surface area being heated. In some cases, the energy can be as much as used in the melter itself. Therefore, the waste gas from combustion in a gas-fired forehearth is not a minor volume of waste gases. These gases, combined with the volatile losses from the glass, are a questionable addition to the plant environment. Even with the direct venting of the forehearth gases to the outside, they can constitute a significant part of the plant's total emissions.

In addition, the vents used in the gas-fired forehearths with these glasses are both a maintenance and glass defect problem. Condensation on the refractory vents builds up, and this material falls back onto the glass surface from these uncovered designs. Some of this condensate melts and corrodes the refractories and the refractory-bearing drip falls into the glass as well.

B. Information on Losses from Forehearths – Expected Effect of Joule-Heated Forehearths

We have already discussed the joule-heated forehearths as "zero-loss systems", as they do not have a fossil fuel combustion system and the attendant waste gases (except in an emergency and start-up). However, we do not want to imply there is a magic bullet that changes chemistry and physics to prevent the ingredients of hot glass from volatilizing from the surface. There are reasons that the all-electric forehearths will see less volatilization from the contained glass, but it will not be zero. The over-riding reason that the all-electric forehearth can claim zero emissions is because the forehearth is sealed, and products of volatilization are kept inside.

1. Reduction in volatilization losses would be logical with Joule-Heating

No flushing of atmosphere by glass combustion – The driving force for volatilization of materials from the glass is to establish an equilibrium concentration of the various species in the gas space above the glass. With the continuous flow of combustion waste gases through the top space and then out, the glass surface never approaches equilibrium with its atmosphere, and volatilization continues at a steady rate.

Colder glass surface with Joule-Heating - Maintaining a glass temperature with flames in the atmosphere will result in a temperature gradient from the flame to the glass surface, and to the bulk glass. With only Joule-heating of the glass, the peak temperatures would be within the glass, and the glass surface and the atmosphere will be colder. In a Joule-heated forehearth, the temperature of the glass surface will be $50 - 75^0$ C lower than in a comparable gas-fired forehearth.

Keeping it sealed completes the job - As we have said, there is no stopping the desire for materials to volatilize from the surface; trying to reach equilibrium in the gas space above the glass. The important part about keeping the forehearth sealed, is that all the outside can be insulated so that all the inner surfaces will be kept hot, minimizing significant condensation of alkali and borate species on the refractories.

2. Evidence in the literature concerning losses from gas-fired forehearths

The amount of loss from forehearths and channels is not well documented in the open literature. Several papers by Ruud Beerkens and his colleagues from TNO have approached this as part of their characterization of the Dutch glass industry. Ruud and Johannes van Limpt, seven years ago at this conference, presented an erudite mathematical development to explain the data on dust emissions for high-loss glasses such as borosilicate C-glasses and E-glasses. They also correlated their predictions with some laboratory data on volatilization from various glasses. While they were addressing the higher temperature regime of the melter, the information can give us guidance for forehearth applications.

Losses from C-Glasses

Figure 13 shows calculated and measured losses from a C-glass. The vertical dashed red lines delineate the range of glass temperatures we would expect in a C-Glass forehearth. From this glass, loss from the glass would be as the species $NaBO_2$ Therefore, this data represents both sodium and boron losses. This species is formed in the glass, so that the water vapor pressure in the atmosphere (the issue of gas-air vs. oxy-fuel) does not play a large role. However, temperature does have a major effect. The lower temperature in a forehearth compared to the temperatures in the melter is significant, giving less loss. But in addition, the 50-75° C lower surface temperature seen with electric, in-glass joule heating, as opposed to gas-firing, would result in further reduction yet. Just looking at a 50^0C reduction in the glass surface from moving away from above-glass burners, we would expect over 30% less volatility loss at the coldest point over the bushings. Where the hot glass enters from the melter, we would expect a 25% reduction.

Another calculation from the TNO paper shows the effect of the velocity of the combustion gases passing over the glass surface. This moving gas removes species that have already evaporated, and drives further evaporation to replace the material swept away. A summary provided to us of this calculation was that losses were proportional to "$v^{0.5}$", if the waste gas flow is laminar to the surface, and proportional to "$v^{0.8}$", if there is turbulent flow at the surface. This shows the major importance of the waste gas flow in the forehearth chamber above the glass in increasing losses from the glass. Therefore, as suggested above, a stagnant air space (or no space at all) in an electric, joule-heated forehearth will result in a significantly lower loss of volatiles as compared to the gas-fired units.

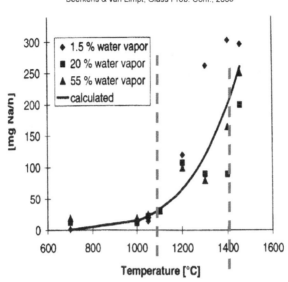

Figure 13
Sodium Release from C-Glass – Laboratory Melts
Beerkens & van Limpt, Glass Prob. Conf., 2000

Losses from E-Glass

We have received requests for proposals for the use of Joule-heating in E-glass forehearths. Losses of volatiles in an E-glass forehearth can also be significant. This is due both to the volatile nature of the boron in the glass, and to the long expanse of heated channels required to supply 60-80 separate Platinum bushings

Again from Beerkens and van Limpt, Figure 14 shows laboratory data on volatile losses of boron species from E-glass. From an E-glass (with almost no alkali), volatility is dependent on water vapor in the atmosphere. The main species from the glass found in the atmosphere will be HBO_2. Thus in Fig. 14 we see the expected steep increase loss of volatiles with temperature, but also a major influence by the partial pressure of water vapor. Therefore, with the dry atmosphere above the joule-heated E-glass forehearth/channel, the boron losses will be much lower.

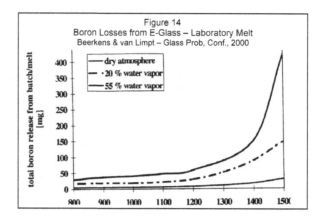

Figure 14
Boron Losses from E-Glass – Laboratory Melt
Beerkens & van Limpt – Glass Prob, Conf., 2000

XI.   ENERGY BALANCE EXAMPLE – C-GLASS

A.   Operating Cost of a Joule-Heated System Can Be Favorable

As discussed before, another benefit offsetting the high capital cost of the joule-heated front-end, in addition to much lower emissions, is the possibility for reduced operating cost. The total energy use in some front end/forehearth systems may be as much as that for the glass melter. As we will see later, use of electricity via Joule-heating in the glass itself can be surprisingly competitive due to a transfer efficiency of over 95% compared to less than 35% from the natural gas burners.

Figure 15 presents a summary of a detailed energy balance for a typical forehearth / front end. This hypothetical example of a C-glass forehearth system has two forming sections and a total length of 210 feet. The energy requirement for each control section of the system was itemized, and then tabulated to show the total energy requirement as if:
  (1) the energy were supplied via direct joule heating within the glass, or instead,
  (2) the energy were supplied by gas-air burners above the glass.

Figure 15							
Heat Balance for Common C-Glass Forehearth System							
Glass	Forming Temp	Forming Sections	Total length	Pull	Status	Total Energy	
						1). Joule-heated	2). Gas-fired
C-Glass	1120°C 2050°F	Two	210 ft 65m	200 ton/day 182 tonne/day	Operating full pull	886 kW	2,953 kW 10,776 scf/hr

Using this energy balance, Figure 16 shows the sensitivity of operating costs to energy pricing. With the market price of natural gas on the left Y-axis and that for electricity on the right, the operating cost per day is shown on the X-axis. Picking a common energy pricing of $7.00 /mscf for natural gas, and $0.05 / kWh for electricity, there wouuold be an $1800/day cost for using gas, but only $1100 for electric. In this particular situation, there would be a significant operating cost advantage to the joule-heated forehearth.

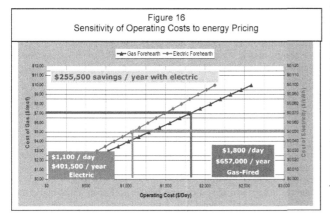

Figure 16
Sensitivity of Operating Costs to energy Pricing

In Figure 17 below, we use the same chart in a slightly different fashion. The particular operating cost of this forehearth will be determined by glass tonnage, temperature, and, of course, the energy source being used and its pricing (let us assume natural gas). Figure 17 shows the market price of electricity that would result in the same operating cost. Any lower price would yield an operating savings.

Figure 17

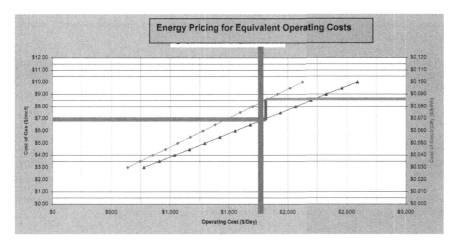

B.   Ultimate Cost of Electricity versus Market Cost

The Department of Energy and the EPA encourages all of us to consider the efficiency of electricity as starting at only 33% (national average for generation efficiency). This would make the ultimate cost of electricity much higher than the 5 – 12 cents/kWh that comes out of our pocket. Unfortunately for this broader point of view; electricity is simply another one of the energy sources for our plants, and if we purchase and use one kW of electricity for joule heating of glass, we can convert over 90% of that kW directly to hot glass.   Our plants are logically forced to make cost-effective choices based on the market price we have to pay.

XII. CAPITAL COSTS OF A FOREHEARTH / FRONT-END SYSTEM – JOULE-HEATED VERSUS GAS

If properly designed, either gas or electrically heated forehearths can deliver glass to the forming process that is thermally homogeneous and at the correct temperature. However, as we have discussed, there is a significant capital cost difference between the two types. With the cost of molybdenum electrodes, water-cooled electrode holders, multiple large transformers, extensive high-wattage cabling, plus the significant labor of installation, the all-electric forehearth has an initial higher capital and installation outlay. This difference can vary from 1.5 to 4 times that of a gas-heated forehearth. There are many choices to be made that affect the ultimate cost.

The number of zones and electrodes, as related to the degree of thermal homogeneity required, is a factor in the total cost. The choice of water-cooled versus dry electrodes can also be significant. The length of the forehearth is also obviously important in determining cost. However, increasing the length by 25% would increase the cost much less than 25%.

Another factor in determining cost is whether the cooling section (alcove) is provided with the very large power inputs that would allow recovery or hot hold automatically with joule heating, as in the system we described above. Some electric forehearths have only the minimal joule heating required at operating conditions, and would require the installation of the emergency gas-firing system to provide recovery and hot hold.

Each installation needs to be carefully considered and designed to provide the required temperature control at the minimum installed cost. KTGSI uses its extensive practical and theoretical experience, working with the customer's input. to optimize designs for both capital and operating cost

XIII. SUMMARY – ELECTRIC JOULE-HEATED FOREHEARTHS SHOULD BE CONSIDERED AT THE PLANNING STAGE

Electric joule-heated forehearths or front-end systems can provide the accuracy and uniformity of temperature needed for forming. The capital cost, however, is a consideration to be weighed against future operating costs.

The high application efficiency, combined with today's volatile and localized energy pricing, may make the daily operating cost of an electric forehearth lower than with natural gas.

Another benefit offsetting the capital cost is the nearly zero level of emissions. The venting of the forehearths into the plant working area, or even outside. is a concern now. It may be regulated soon. The electric forehearth system offers a answer.

Low NOx COMBUSTION IN REGENERATIVE GLASS FURNACES

Ruud Beerkens and Hans van Limpt
TNO Science & Industry
Eindhoven, The Netherlands

SUMMARY

Primary measures, to avoid formation of mainly thermal NOx (NO formation from a reaction between nitrogen and oxygen in the core of a flame at high temperatures) have been applied since the eighties of the twentieth century in glass melting furnaces. The main modifications have been:

a). changes in burner types to allow slower mixing between fuel and oxidant (air or oxygen) and to promote soot formation increasing the radiation intensity and cooling of the flames,

b). improved control of the actual air or oxygen excess to approximate stoichiometric combustion,

c). a decrease of the number of burners per port, and

d). changes in burner settings, such as burner angle and the injection nozzle size, that determine gas/fuel injection velocity and mixing rates.

However, in many cases, a decrease of NOx concentration levels below about 900-1200 mg NO_2/Nm^3 (typical 3 to 4 pounds NOx per ton container glass melt), resulted in increased CO (carbon monoxide) concentration levels in the exhaust of the furnace and top sections of regenerator chambers and in several cases dust formation and SOx emissions increased as well.

In order to obtain lower NOx formation levels far below 1000 mg/Nm^3 in regenerative fired glass furnaces, combustion chamber design and burner port modifications seem to be required. CFD (Computational Fluid Dynamic) modeling is an important tool to find furnace designs in combination with a burner system resulting in lower NOx formation rates, but at the same time, avoiding an increase in CO, SOx emissions and increased evaporation from the melt that may lead to extra particulate emissions.

1. INTRODUCTION

The production of glass in fossil-fuel fired furnaces is always associated with emissions of NOx: NO and NO_2. NOx emissions by traffic and industry appear to be a precursor of acid gas formation (HNO_3) and NOx will increase ozone formation in the upper troposphere and lower stratosphere, see figure 1.

The main sequence of reactions which creates ozone in the free troposphere is through the photolytic destruction of NO_2 [1]:

$$NO_2 \Rightarrow NO + O \text{ (driven by UV radiation)} \tag{1}$$

and

$$O + O_2 \Rightarrow O_3 \text{ (ozone)} \tag{2}$$

187

Organic contaminants in the atmosphere can oxidize NO into NO_2, such as alkyl-peroxy radicals (RO_2):

$$RO_2 + NO \Rightarrow NO_2 + RO \qquad (3)$$

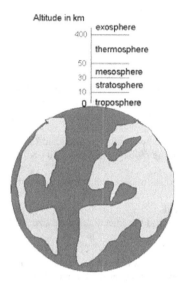

Altitude in km

400 — exosphere

thermosphere

50 — mesosphere
30
stratosphere
10
0 — troposphere

Figure 1   Altitude levels in the earth atmosphere.

Thus, NO may be transformed into NO_2 in the atmosphere containing organic compounds and hydroxyls ($RH + OH + O_2 \Rightarrow RO_2 + H_2O$). This NO_2 leads to ozone formation (reaction 1 and 2) or acid rains [2, 3].

In the combustion space of glass furnaces, most of the nitrogen oxide is in the NO form [4]. The main mechanisms of NOx formation (the term NOx represents all nitrogen oxide compounds, the NOx emission often expressed in mg per Nm^3 flue gas or kg/ton molten glass and calculated as NO_2) are fourfold:

a. NITRATE decomposition (KNO_3 or $NaNO_3$) in the batch blanket: 600-900 °C

$2NaNO_3 \Rightarrow 2NO_2 + Na_2O + 1/2O_2$ at about 600-900 °C
Part of this NO_2 dissociates into: $NO + 1/2O_2$

b FUEL BOUND nitrogen (mainly HCN-components in oil, coal) [5]: nitrogen atoms bonded to organic molecules react with oxygen, forming NOx

c. PROMPT NOx near flame zone (very rapid):
PROMPT NOx is caused by the breakdown of CH portions [5-7] of methane and other hydrocarbons in the fuel and their rapid subsequent combination with nitrogen in the air. Hydrocarbon radicals reacting first with molecular N_2 are forming amines & cyano in the flame front. These amines or cyano compounds will further oxidize and can form NO.

Prompt NO formation is only important in the first parts (front part) of the flame, when hydrocarbon components are still present.

d. THERMAL NOx: according to the Zeldovich mechanism (strongly temperature dependent) [8]:
NOx formation takes place, when at the same moment and position: O_2, N_2 and high temperature (T > 1700 K) co-exist. The main reactions involving reactive species and forming the so-called thermal NO are:

$$O + N_2 \Rightarrow NO + N$$
$$N + O_2 \Rightarrow NO + O$$
$$N + OH \Rightarrow NO + H$$

For natural gas firing, the mechanism number 2 is not significant and in most cases of glass furnace combustion processes, mechanism 3 (prompt-NOx) is a minor source for NO or NO_2 emissions.

Thermal NOx is the most important mechanism for many glass furnaces. Whenever high temperatures (>>1800 K), free oxygen, free nitrogen co-exist for a certain time, NOx formation takes place within this time period. But, at certain conditions, mostly fuel rich and oxygen lean circumstances, the NO and NO_2 species may react with the hydrocarbons and form N_2 (plus water or CO and CO_2) [9], thus the NOx concentrations are reduced.
In order to reduce the emissions of NOx in the glass industry, primary measures for air fired combustion processes in glass furnaces or oxygen combustion, as well as secondary measures can be applied. In the first case, the combustion process has been modified (e.g. burner settings, fuel-air ratio control, delayed mixing, geometric adaptations of the combustion chamber and burner ports etc). In the second case, end-of-pipe measures like ammonia injection in the combustion gases with or without catalysts (SCR and SNCR) and fuel injection in the hot combustion gases (3R process) are applied.

In several literature references, NOx reduction techniques applicable [11-18] for glass furnaces have been reported. In many cases, the results of combustion tests in pilot furnaces and glass furnaces are reported, but other possibilities, which are still in its infancy, are discussed also. Often the evaluation and identification of (often non-desired) side-effects observed during the application of the different primary and secondary NOx reducing measures, like:

• Higher evaporation rates from the melt or batch blanket in the glass furnace,
• Higher CO concentrations,
• Interaction of combustion gases with refractory materials,
• Higher energy consumption or
• Emissions of other compounds like NH_3,
are not mentioned in literature, especially when authors want to promote a certain technology.

Other aspects regarding NOx emission reduction for glass furnaces, such as:
• Effect furnace (combustion chamber) design on NOx emissions;
• Effect of side-of-the port versus under-port firing on achievable NOx emissions;
• Effect of number of burners (for same total fuel input) on NOx formation;

- Effect of so-called FLOX (flameless oxidation, achieved by intense flue gas re-circulation and dilution of the flame) combustion on NOx emissions and furnace behavior,

have hardly been investigated yet.

Because of several non-desired side effects that are possible during the application of low-NOx measures and the incomplete information on the effects of burner positions and furnace design on NOx formation, the project: 'low NOx regenerative furnaces' has been initialized. The objective of the project is to design modifications (especially in burner ports and combustion chamber), burner selection or burner types (new burner concepts) and combustion control for cross-fired and end-port fired regenerative glass furnaces in order to obtain NOx emissions in the range of 600-750 mg/m$_n^3$, but without the non-desired side effects, see section 3 of this paper.

This paper will discuss the possibility of reducing NOx emissions of air-fired glass furnaces and some results from detailed industrial tests and experiences are shown

2. PRIMARY MEASURES

The most important primary measures for NOx emission reduction are listed below:

1. Selection of burner type, that allows a reduction of the fuel injection velocity in order to reduce the mixing between the fuel and air (or oxygen) and to increase the residence times of the flame in the furnace. Especially at natural gas injection velocities far above 100 m/s, NOx formation rates are high, because of fast mixing of air-gas resulting in high temperature flame cores.

2. Selection of burner types that create separate fuel flows by using central and annular nozzle types. The central fuel jet is shielded, from direct contact with preheated air, by an outer fuel stream. The central fuel flow, heated by radiation and not reacting with oxygen will create extra soot. This soot with a high emission of NIR radiation will cool down the flame and increases the radiation heat transfer to and into the melt [10].

   For instance, oil burners have been developed with a narrow-shaped size distribution of the injected oil droplets [11]. An optimum size of oil droplets avoids a rapid combustion and very high flame temperatures occurring in case of fine droplets and prevents the incomplete combustion of coarse droplets. Due to the narrow size range of the oil droplets of STG DeNOx oil burners early combustion is avoided and a long (but not too long) low temperature flame is obtained with a wide area for energy release to melt and batch. STG reported for a 6-port float glass furnace equipped with STG DeNOx oil burners in combination with air/fuel ratio control, based on oxygen sensors, that NOx emission levels as low as 700 mg/Nm^3 could be achieved. However, the possibility of reaching these levels depend, on many parameters, including furnace design and acceptable flue gas conditions.

3. Selection of burner types that allow a variation of the vertical and horizontal angle. The angle between the burner and the flow direction of incoming air or oxygen will effect the mixing rate. A smaller contact air-fuel angle will often lead to delayed mixing and longer, but also more luminescent (more soot), wider flames.

4. The air or oxygen excess: it is possible to create flames with very low NOx formation by under-stoichiometric combustion. But, this is not acceptable in most glass furnaces as demonstrated in section 5 of this paper. Important is the sealing of the burner blocks to avoid open joints between burner and blocks. Cold air penetrating into the

furnace through open joints or open doghouse constructions will increase energy consumption, and open joints close to the burner areas can lead to increased NOx formation levels. The prevention of this cold air infiltration can often be achieved by (very) simple measures. The level to which excess air can be reduced in a cross fired regenerative furnace may be limited, because many cross fired furnaces do not always have the possibility to control the air-fuel ratio per burner port. STG [11] developed a system to control the air flow per regenerator chamber segment and burner port by the application of air curtains regulating the pressure and combustion air flows in the regenerators.

Abbasi [12, 13] showed in several excellent papers in the eighties, correlations between air excess, air preheat temperature and fuel-air mixing on NOx formation, measured in a pilot-scale arrangement simulating burner ports of glass furnaces and under-port, side-of-the port and over-port firing. The correlations were checked and confirmed on industrial furnaces [13].

At that time, NOx emissions on existing regenerative furnaces could be decreased from about 1600-1900 mg/Nm^3 (8 % O_2, dry) down to values below 1000 mg/Nm^3 for container glass furnaces, using burners with variable nozzle-orifice sizes. For each 100 °C increased air preheating, the NOx emissions (other parameters kept constant) increased by 37 %. An increase of the air excess by 5 % (5 % from the stoichiometric level) increased the NOx emissions by 20-30 %. Burner angle, port design and positions of burner are other important parameters. The NOx-emission level reduction depends of course on the starting (initial) situation and a reduction from 2500 mg/Nm^3 down to levels below 1200 mg/Nm^3 are more easily achieved than below 1000-1200 mg/Nm^3 without non-desired side-effects. Abbasi concluded that burners in the side-of-the-port arrangement with an angle of 90° relative to the air flow gave higher NOx emissions compared to under-port burners or side-of-the port burners with lower angles. However, in the glass industry it appeared that side-of-the port burners (90° angle relative to the air flow) in combination with a baffle or step in the burner port, appeared to be effective to obtain low NOx emissions [14]. Neff [15] stated that the combination of air velocity and burner angle determines the mixing rate. In cross fired furnaces with multiple burner ports at each side, the (preheated) air velocity in the ports is lower compared to end-port fired furnaces in most cases. Side-of-the-port firing at 45° burner angle in cross fired furnaces (relatively low air velocity in most cases) will lead to medium mixing rates and medium level NOx concentrations. In an end-port furnace with high combustion air velocities in the burner port, application of such side-of-the port burners could lead to increased NOx emissions compared to under-port firing.

Side-of-the port firing, using burners lined at 90° relative to the combustion air flow in the port in combination with an air baffle [14] to by-pass part of the air and operating at low gas injection velocities may lead to lower mixing intensity and therefore lower NOx emission levels. Neff [15] showed the possibilities to pre-crack a part (about 25 %) of the natural gas in separate crackers using oxygen to burn the gas partially and to create soot. The pre-cracked gas and the main natural gas flow (75 %) are combined as a soot rich gas mixture at about 290 °C. This mixture is used as the fuel for the glass furnace burners. This technique can lead to more intense flames with high radiation and better heat transfer and lower NOx emissions and lower exhaust temperatures (by increased heat transfer to the melt). Thus, it will lead to lower air preheat temperatures. These factors will result in lower NOx emissions. This technology, developed in the eighties of the twentieth century however is hardly applied in the glass industry.

Figure 2 shows, the NOx concentration in a furnace atmosphere, depending on oxygen content in exhaust gas (or air excess) and temperature for air-firing and oxygen-firing, according to thermodynamic equilibrium conditions. This figure is not applicable to glass furnaces as such, because of the limited validity of the assumptions of thermodynamic equilibrium, but it shows the dependencies and trends.

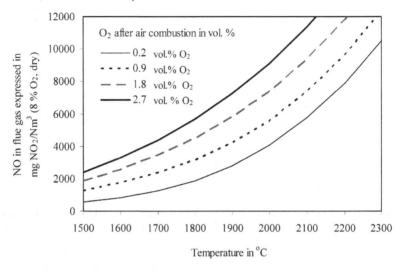

Figure 2   *NOx concentrations in flue gas from combustion atmosphere according to thermodynamic equilibrium conditions and depending on combustion temperature and air excess (O_2 content after combustion).*

5. Elimination of the cold primary air in natural gas burners, the primary air is often used to increase the gas injection velocity, but this cold air leads to extra NOx formation in the root of the flame and increased energy consumption. For example, 3 % cold air (3 % from total combustion air) infiltration increases the energy consumption of a regenerative furnace with 1.5-2 %. ;

6. Design changes in the combustion chamber: decreasing the slope of the burner port crown (see Sieger [16]) to values less than about 23-27°, increasing the burner port opening to lower the combustion air velocities (< 10 m/s), and increasing the height of the crown (distance crown – melt) to achieve longer residence times of the combustion gases to complete combustion, even for cases with very delayed mixing of fuel and oxidant.

7. A change of the positions of the burners or the number of burners per port [17, 18] or application of an extra pre-combustion burner upstream the air in the burner port [19] (cascade burner systems). However, cascade burner systems lead to only limited additional reduction of NOx emissions in cases with already rather low NOx emission levels (about 700 mg/Nm³), see Tackels [16]

8. Oscillating combustion or pulsed combustion (tested in oxygen fired furnaces [21]). But, to authors' knowledge, there is hardly any experience with such burners in industrial glass furnaces.

Several of these measures have been applied to end-port fired regenerative container glass furnaces. Case study: The initial NOx concentrations in the flue gases of a natural gas fired furnace were in the range 2500-2800 mg/Nm³ with an air number of about 1.17. Just, blocking the primary air inlet resulted in 10% reduction of NOx emissions and 6 % reduction in the air number caused a NOx concentration reduction down to 1850 mg/Nm³. After installing burners with adjustable burner angle and nozzle size (to adjust the gas injection velocity and to keep natural gas velocity below 90 m/s), maintaining the air excess below 5-7 % and keeping CO content still below 250 ppm (in burner port), the NOx concentrations could be lowered down to 1200 mg/Nm³. Further decreases of the NOx concentrations appeared to be possible, but only at fast increasing CO emissions levels.

3. DECREASING AIR EXCESS IN END-PORT FIRED FURNACE

At the end of the furnace campaign of an end-port fired container glass furnace in Europe, there was the opportunity to adjust the combustion process of this furnace to an extent that was not desired at normal production situations. Thus, the effective air number could be decreased down to values close to 1 (almost no oxygen in the exhaust gas). The existing gas burners (with two gas jets per burner: a core flow and an outer jet surrounding the core jet with adjustable burner angle and burner nozzle), 3 burners per port were used with average gas injection velocities estimated on about 60-85 m/s. In the top of both regenerator chambers, the concentrations of particulate, SO_2, NOx, CO and O_2 have been measured. Objective of these trials was to investigate the effect of low air numbers on NOx emissions, but also on other emissions that could be affected by changing the combustion process.

The air number was decreased from about 1.12 to almost 1.01. Figure 3 shows the results of the concentration measurements in the top of one of the regenerator chambers. An almost linear decrease of the NOx concentration with decreasing residual oxygen content could be observed. However, in this case, below oxygen contents of about 0.6-0.8 vol. %, the CO content of the exhaust gases entering the regenerator increased very rapidly. Remarkable is that also the particulate and SO_2 concentrations increased. Obviously, a reduction of air excess and the chance of reduced conditions in the furnace atmosphere will increase the evaporation of glass components and decomposition of sulfates from the batch. The condensation of evaporated glass components (sodium species and sulfur species) in the cooling flue gases leads to the increased particulate formation.

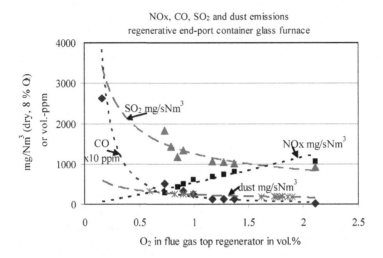

Figure 3   *Measured concentrations of exhaust gas components in the top of the regenerator of an end-port fired container glass furnace, depending on the oxygen content in the exhaust gas (trials with reduction of the applied air number)*

The figure shows that very low NOx emission levels can be achieved, but the same figure shows that in this case, these low NOx concentrations are associated with very high CO, SO_2 and particulate levels. High CO levels entering the regenerator checkers or reduced vapors may react with certain refractory types (especially type containing iron or calcium) and in case of such reducing conditions, refractory material selection becomes very important [22]. Beerkens and Van Limpt [23] showed that reduced firing (fuel rich) conditions and reduced combustion atmospheres increases the evaporation of sodium due to the reaction at the glass melt surface;

$$Na_2O \text{ (glass melt)} + CO \rightarrow 2Na(vapor) + CO_2$$

Beerkens [24] showed that reduced gases in direct contact with the batch blanket stimulate sulfate decomposition:

$$Na_2SO_4 \text{ (batch)} + CO \rightarrow Na_2O \text{ (batch/melt)} + CO_2 + SO_2$$

4. FURNACE DESIGN MODIFICATIONS AND COMBUSTION PROCESS OPTIMIZATION

Most studies at existing regenerative glass furnaces showed that NOx emissions can often be reduced by primary measures, but below a certain NOx concentration level, other emissions, see section 3 may increase. This is often caused by the delayed mixing of the fuel with air, leading to a delayed combustion process, and consequently local reduced conditions. The chance on non-completed combustion, because of these lazy flames and limited residence times of the combustion gases in the combustion space of the glass melt furnace, limits the possibility of reducing the air excess to near stoichiometric conditions. An increased residence time and prevention of a direct contact of reducing parts of the flames with the batch (reduction of sulfate containing batch causes SO_2-emissions) and melt (decreasing evaporation of alkali compounds from the melt) can be achieved by changes in the burner port and combustion chamber designs. Figure 4a and 4b show for two different combustion and burner port designs the trend in NOx formation in the flames.

The calculations (results for NOx shown in figure 4a and 4b) show that the NOx contents decrease by about 15-20 % by increasing the height of the combustion chamber with 4 inches. The modeling results showed even a reduction in the CO concentrations despite the lower NOx contents when increasing the height of the combustion chamber.

A higher (tall) crown generally results in an increased re-circulation of flue gases or allows a longer residence time of the combustion gases in the furnaces, this enables

a). delayed mixing and
b). still complete combustion,
c). lower air excess and
d). consequently lower NOx emissions.

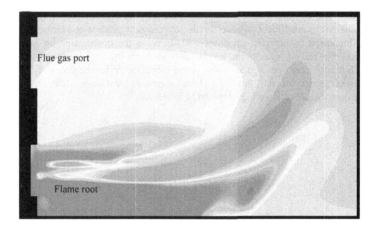

Figure 4a   *NOx trends determined by CFD plus NOx modeling in combustion process of a regenerative fired glass furnace for a base-case design . The NOx contents in the furnace atmosphere is shown by color contours at a horizontal cross section of the combustion chamber at the level of the burners in an end-port fired regenerative furnace. NOx levels increases from blue to green to yellow to red.*

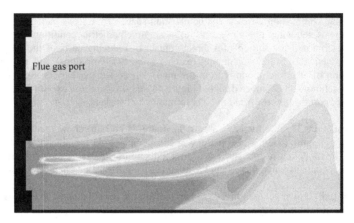

Flue gas port

Figure 4b   NOx trends in combustion process of a modified regenerative fired glass furnace with increased crown height (4 inch higher combustion space compared to base case in figure 4a).

Greczmiel and Scherer [26] reported a reduction of NOx emissions from 1440 down to 1200 mg/Nm^3 due to a change in furnace design applying a higher superstructure. An increased residence time by increased combustion space also results in thicker flame layers (more radiation) and more time available for the combustion process to release heat to melt and batch. Increased burner port openings will decrease the combustion air velocity and thus decreases the mixing rate of the combustion air with the fuel injected in the combustion space, but such port openings will also cause increased residence times of the flames in the combustion space of the furnace. These increased residence times, allowing delayed combustion, enables a lowering of the NOx formation.

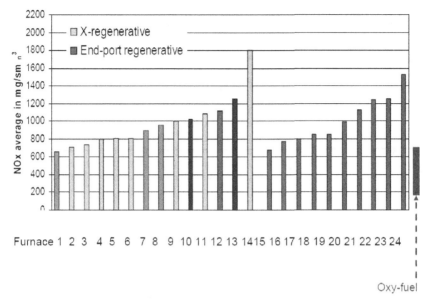

Figure 5   *Measured NOx emissions of several European glass furnaces.*
The bar at the right hand side shows the equivalent NOx emission range of
oxygen-fired glass furnaces (calculated as equivalent for typical air-fired flue
gas volume flows)

The range of NOx emissions for a number of modern cross fired regenerative glass furnaces and end-port fired regenerative furnaces is presented in figure 5. The typical NOx emission range for oxygen-fired furnaces, after taking into account the change in flue gas volume is also presented. Difference between furnaces are mainly caused by different furnace temperatures, air excess differences, design changes and type of burners/burner settings.

5. TRIALS IN PILOT TESTS FACILITIES

Few pilot test facilities have been applied in Germany and the USA to study the effect of air preheat temperature, burner types, gas injection velocities, air number, burner angles and staged fuel injection (such as cascade burners [19]) on heat transfer and NOx concentration in the resulting exhaust gases [12, 13, 17, 18]. Abbasi et al [12] derived already in 1985 correlations between the burner angle, air preheat temperature and natural gas velocities and NOx formation by extensive experiments. Scherello [17] used CFD calculations and a test facility at GWI (Gas Wärme Institut in Essen, Germany) to investigate flames and pollutant (CO, NO) formation as well as heat transfer. Scherello used pipe-in-pipe under-port burners with an inner jet of natural gas and second flow of gas (surrounding the central jet) and measured NOx concentrations. Large gas velocity differences between the two jets resulted in strong turbulent mixing and high reaction intensity in the flames and consequently in high NOx formation rates. Secondary natural gas injection in an end-port fired furnace from the sidewall in combination with fuel lean primary flames from under-port burners showed a decrease of 35-45 % in NOx emissions compared to the reference case with only under-port burners and no secondary fuel injection. High momentum burners not always lead to increased NOx emissions in glass furnaces according to Scherello. High momentum burners may lead to intensive mixing of preheated air and fuel and gives very hot flame cores, but on the other hand the increased recirculation of the flue gas inside the furnace dilutes the flame root and may decrease the oxygen content in the flame. So far it is not well known in which way an increased gas injection velocity will effectively lead to lower NOx emissions in glass furnaces. Scherello et al. [18] described their test facility for trials with typical glass furnace burner systems using preheated air up to 1350 °C and one under-port positioned burner (total power up to about 1-1.5 MW) and a combustion air inlet port (above the burner). Such test facilities can be used to find dependencies between burner type and burner setting and NOx emissions and heat transfer and are used to validate CFD models (Computational Fluid Dynamic simulation models describing gas flows, mixing and combustion). However, results from such tests may differ strongly from measurements in industrial glass furnaces due to the large effect of the dimensions and design of the combustion space on recirculation, flame length, mixing, flame temperatures and NOx-formation rates.

The vertical burner angle and gas velocities can be changed on-the-run by adjustable burners types. In their tests with different burner angles and gas velocities (low and high momentum), the NOx concentrations in the flue gases increased in the case of a tested burner type from a range of 450-600 mg/Nm^3 (8 % O_2, dry) at air preheating of 1100 °C up to a range of 980-1500 mg/Nm^3 at 1300 °C preheated air. The most optimum burner angle and gas velocity with respect to a minimum NOx emission level, the heat transfer profile, and flame contours depend on the applied burner type. UV visualization techniques (light intensified CCD camera in combination with small band UV-filter) are used at GWI to observe and to measure the reactive part (with OH radicals) of the flames. OH radicals in the flames react with CO to complete combustion. Tuning of flames and flame length adjustment require such techniques, because not the complete reactive part of the flame can be observed as being radiating in the visual spectrum. Part of the flame mainly radiates UV.

6. CONCLUSIONS
Primary measures have been successfully applied to decrease the NOx emissions of regenerative and air fired glass melting furnaces from levels far above 2000 mg/Nm^3 to levels between 800-1200 mg/Nm^3. A further reduction of the NOx emissions of air-fired regenerative furnaces by combustion measures is often associated with increased SOx and higher particulate emissions (sodium), and flue gases containing CO and other reduced species flowing through the regenerators. A further step in decreasing the NOx emissions of glass furnaces can be achieved by modifications in the combustion chamber design and burner port geometry. Enlarged combustion spaces in combination with adjustments in burner positions (distance from melt) and burner settings will probably enable extra decreases in NOx formation rates without running into problems of reduced flue gases with too high CO contents entering the regenerators. Development of regenerator checker refractories [22] that resist increased alkali vapour concentrations, reduced flue gases and withstands reduced salts depositing from the flue gases on the refractory surface enlarge the operating window and may allow an almost stoichiometric combustion process in regenerative fired glass furnace with NOx levels far below 1000 mg/Nm^3.

Reported NOx emissions of glass furnaces after applying primary measures have limited sense without reporting correctly measured CO concentration levels of the flue gases entering the regenerator. The combination of low NOx and CO contents is the challenge in the application of primary measures.

ACKNOWLEDGEMENT:
CFD calculations of the combustion processes and NOx formation in glass furnaces have been performed by Dr. Adriaan Lankhorst and Ing. Andries Habraken from TNO Science and Industry.

REFERENCES

[1]   Crutzen, P.J.: The influence of nitrogen oxides on atmospheric ozone content. Q.J.R. Meteorol. Soc. **96** (1970) pp. 320-325

[2]   Grove, M.; Sturm, W.: NOx Abatement System: Using Molecular Sieve Catalyst Modules for a Glass Melting Furnace. Ceram. Eng. Sci. Proc. **10** [3-4] (1989) pp. 325-337

[3]   Acid   Rain,   Environmental   Protection   Agency   (EPA),   August   6th,   2002 http://www.policyalmanac.org/environment/archive/acid_rain.shtml

[4]   Kircher, U.: Emissionen von Glasschmelzöfen-Heutiger Stand. Glastech. Ber. **58** (1985) nr. 12., pp. 321-330

[5]   Koster, C.L.: Modelling of NOx Formation in a High Temperature Gas-Fired Furnace. PhD thesis University of Technology (Technische Universiteit) Delft, The Netherlands Delft University Press (1993) ISBN 90-6275-883-5

[6]   Fenimore, C.P.: Formation of nitric oxide in premixed hydrocarbon flames. Proc. 13th Int. Symposium on Combustion (1971), pp. 373-380

[7]   Miller, J.A. and Bowman, C.T.: Mechanism and modeling of nitrogen chemistry on combustion. Proc. Energy Combust. Sci. (1989), **15**, pp. 287-338

[8]   Zel'dovich, J.: The oxidation of nitrogen in combustion and explosions. Acta Physicochimica URSS, XXI (4) (1946) pp. 577-628

[9]   Schulver, I.N.W.; Quirk, R.: Pilkington 3R Technology.: An up-date. Ceram. Eng. Sci. **18** (1997) [1] , pp. 60-65

[10]   Faber, A.J.; Van Nijnatten, P.A.: Determination of the optical properties of glass melts. Kurzreferate of the 78. Glastechnische Tagung 7.-9. June 2004, Deutsche Glastechnische Gesellschaft Nürnberg, pp. 119-123

[11]   Hemmann, P.: Advanced combustion control-the basis for NOx reduction and energy saving in glass tank furnaces. Glass Sci. Technol. **77** (2004), no. 6, pp. 306-311

[12]   Abbasi, A.: Khinkis, M.J.; Fleming, D.K.; Kurzynske, F.R.: Reduced NOx emissions from gas-fired glass melters. Gas Wärme International (1985) nr. 8, pp. 325-329

[13]   Abbasi, A.H.; Fleming, D.K.: Combustion Modifications for Control of NOx Emissions From Glass Melting Furnaces. Ceram. Eng. Sci. Proc. (1988) **9** [3-4] pp. 168-177

[14]   Barklage, H.: Erfahrungen mit Primärmaßnahmen zur Verbesserung der Umweltsituation. Fachausschuß VI of the DGG Germany, 14. October 2004, Würzburg

[15]   Neff, G.: Reduction of NOx emissions by burner application and operational techniques. Glass Technol. (1990) **31**, no. 2., pp. 37-41

[16]   Sieger, W.: Feuerungstechnik in Glasschmelzwannen. DGG Fachausschuss II: Ofenbau und Warmewirtschaft – Feuerfeste Baustoffe (1987) 7. April 1987 Würzburg

[17]   Scherello, A.; Giese, A: Entwicklung neuer Brennersysteme für Glasschmelzwannen mit regenerativer Luftvorwärmung. Fachausschußsitzung Deutsche Glastechnische Gesellschaft (DGG), Bad Münder, 18. März 2005

[18]   Scherello, A.; Flamme, M.; Kremer, H.: NOx Reduction and Improvement of Heat Transfer for Glass Melting Furnaces with Regenerative Air Preheating

[19]   Becher, J.; Wagner, M.: SORG second generation cascade heating system for reducing NOx, including additional primary measures. International Glass Journal (2000) no. 109, pp. 40-43

[20]   Tackels, G.: Reducing NOx emissions in glass manufacturing. Glass Machinery Plants & Accessories (2002) no. 3. pp. 105-112

[21]   Joshi, M.; Borders, H.; Charon, O.; Legiret, T.; Zucchelli, P.: Optimum oxyfuel melter performance: oscillating combustion. International Glass Journal (2000) no. 109, pp. 44-51

[22]   Schmalenbach, B.; Weichert, T.; Postrach, S., Heilemann, G.; Lynker, A.; Gelbmann, G.: New Solutions for Checkers Working under Oxidizing and Reducing Conditions. 67. Conference on Glass Problems (2006) 31. October – 1 November, Ohio State University, Columbus OH

[23]   Beerkens, R.G.C; Van Limpt, J.A.C.: Evaporation in industrial glass melt furnaces", Journal of Glass Science and Technology, **74** [(9] 245-257 (2001)

[24]   Beerkens, R..: Sulphur Chemistry and Sulphate Fining and Foaming of Glass Melts. Glass Technol.: Eur. J. Glass Sci. Technol. A, **42** (2007) no. 1, pp. 41-52

[25]   Lankhorst, A.M.: Simulation of Glass Melting Processes at TNO Glass Group Benchmark, Int. Magazine for Engineering Designers & Analysts, April 2007, pp. 23 - 29

[26]   Greczmiel, M; Scherer, V.: Reduzierung der NOx-Emission einer Spezialglaswanne durch primäre und sekundäre Maßnahmen-Erfolge und Irrwege. Fachausschuss VI Umweltschutz (Technical Committee Environmental Protection) Deutsche Glastechnische Gesellschaft (DGG) 15. October 1997, Würzburg, Germany

ADVANCED CLEANFIRE® HRi™ OXY-FUEL BOOSTING APPLICATION LOWERS EMISSIONS AND REDUCES FUEL CONSUMPTION

Michael Habel, Kevin Lievre, Julian Inskip, Jan Viduna, and Richard Huang
Air Products
Allentown, PA

ABSTRACT
Oxy-Fuel boosting is a process technology in which a set of oxy-fuel burners is used to enhance performance of an air-fuel fired melter. This technology has been used primarily to increase glass production rates and to increase glass quality. It has also helped glassmakers to become familiar with the operation of an oxy-fuel system prior to committing to convert their glass melting operation over to complete oxy-fuel firing.

Air Products has recently completed development of a new Cleanfire® HRi™ Advanced Oxy-Fuel Boost Burner that is specifically designed for boosting applications. This new state-of-the-art burner was designed and developed making use of the company's in-house glass team resources: combustion labs, *Combustion Centre of Excellence*, cross-industry knowledge of industrial combustion processes, computational fluid dynamic modeling expertise, and specific knowledge of glassmaker's melting requirements. The technology was designed to further improve the production rate and glass quality benefits that have driven most past installations, and to significantly impact emissions and fuel usage. Recent fossil fuel price increases and ever-tightening environmental regulations have increased the economic importance of preserving fuel and reducing emissions. Several glass manufacturers have put the technology to the test and report very positive results. This paper will provide a review of the technology and how it has positively influenced glass melting.

OXY-FUEL BOOSTING
Business decisions today are primarily driven by the investor or stock holder. Each year Earnings per Share and Investment Value must grow in order to maintain investor interest and monetary growth. There are many strategies that a companies' management team can embrace to improve the performance of their company. One of the most fundamental and easy to understand is the concept of working assets harder - that is improving the return on capital investment via greater utilization. This paper updates the industry on oxy-fuel boosting technology, a technique designed to make greater use of existing glass melting furnace assets.

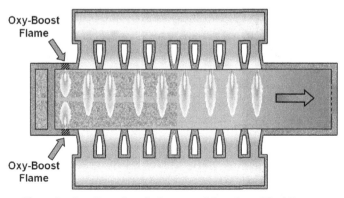

Figure 1 – Oxy-Boost Installation on an 8 Port Cross-Fired Furnace

Oxy-Boost
Flame

Oxy-Boost
Flame

Having introduced the oxy-fuel boosting application to the glass industry, and now with a reference list comprised of installations over a 20-year period, Air Products' glass team is fully versed in techniques that result in a system optimized for each individual furnace. The basic application entails installing discrete oxy-fuel burners into a glass furnace to supplement the existing air-fuel firing system - see Figure 1. To date the oxy-fuel boosting application has been installed on more than 75 furnaces by Air-Products worldwide. The oxy-boost firing system provides the glassmaker with a low capital, short-payback retrofit that can be installed while in production, and with little-to-no disruption to the glassmaking process. As such, this application is a very attractive tool for maximizing the quantity of saleable product. Equally important, because glass furnaces are so capital intensive, glassmakers cannot afford to make mistakes that might shorten the furnace campaign. When installed by a diligent and experienced team possessing glass melting knowledge and burner technology expertise, it has been demonstrated that a multitude of benefits can be achieved with oxy-fuel boosting while avoiding adverse effects on immediate production and melter refractory. As testament to the technology, many operators are using on-site generated oxygen. This mode of supply is only adopted if the technology is intended as a long-term feature of the furnace operation.

Generally oxy-boost can be said to have the following benefits:

- increase the production rate: 5-30% pull rate increase for healthy furnaces (more possible when compared to a crippled furnace)
- extend the furnace life: reduce carryover into regenerators, keep furnace operating with failed heat recovery (failed regenerators, recuperators)
- improve quality: reduce bubbles and inclusions
- improve efficiency: 2-10% fuel savings on a healthy sideport furnace (more for other designs and for crippled furnaces)
- reduce emissions: 5-20% (NOx, CO, CO_2, particulate matter)

A recent trend is the general adoption of boost within the European float industry, with more than 15% of furnaces currently operating with the oxy-fuel boosting application.

The equipment required to install and operate an oxy-boost installation is detailed in the following schematic:

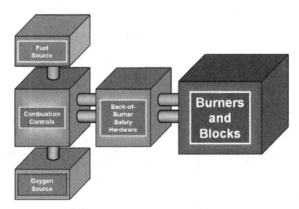

Figure 2 Equipment Schematic

Discussing each element of the installation in turn:

Fuel Source

Any common fuel source can be used with the oxygen-boost system. Systems are in operation firing liquefied petroleum gas (LPG), natural gas, light fuel oil (LFO), and heavy fuel oil (HFO). Essentially, if it has a calorific value it will burn with oxygen and can be successfully used for oxy-fuel boosting when accompanied with technology designed to properly control its introduction.

Oxygen Source

Liquid oxygen (LOX) is most often used to supply the oxy-boost system, at least initially. Road tanker delivered, liquid oxygen is stored in a cryogenic storage vessel at the glass factory. The heavily insulated cryogenic storage tank, holds the oxygen in the liquid state at about -300°F (-186°C). The oxygen is vaporized using either ambient air vaporizers or warm water\steam vaporizers. Typical storage pressures range up to about 350 psig (24 Barg).

Alternatively, the oxygen may be generated at the glass factory. At float glass facilities, this on-site generated oxygen can be a by-product of the nitrogen production, which is used to inert the float bath. In glass factories that do not produce float glass, the oxygen required for boosting may be generated by a dedicated vacuum swing adsorption (VSA) plant... sometimes also referred to as a vacuum-pressure swing adsorption plant (VPSA).

Control System

The function of the combustion control system is to safely deliver, precise amounts of fuel and oxygen to the boost burners. The fuel flow rate is set by the operator and the oxygen flow rate is determined by the stoichiometric oxygen requirement of the fuel. The oxygen flow is metered in a precise and often automatic ratio to the fuel flow. The control system continuously monitors supply parameters, to ensure that if unsafe conditions occur, the system alarms and achieves a safe state. Typically, pressure and temperature corrected mass flow is measured and controlled for both the fuel and oxidant, and a separate control loop is used for each burner. Figure 3 below depicts a typical control system for use with low pressure on-site generated oxygen and natural gas. Appropriate distribution pipework must be installed from the oxygen and fuel sources to each burner.

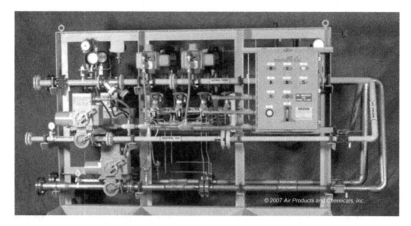

Figure 3 – Flow Control System for Oxy-Fuel Boosting

Back-of-Burner Safety Hardware

The back-of-burner safety hardware includes local, manually operated valves, to ensure that fuel and oxygen can be locally shut off during burner installation and removal. A set of non-return valves (also commonly called check valves) are installed to prevent the risk of either fuel or oxidant travelling back into the other fluid's distribution pipework. Finally a set of flexible hoses are installed. For safety reasons, these hoses are as short as possible, suitable for the maximum allowable pressure, suitable for high temperature, and mounted to avoid distortion, whiplash or accidental damage. In the case of oxygen components, all materials are selected to be compatible, safe for use with oxygen, cleaned for oxygen service, and installed with special precautions followed for equipment to be used in oxygen service.

Burners and Blocks

All of the preceding elements of the system, provided they are well engineered have very little impact on the performance of the system. Just as in automobiles, it is in the combustion chamber where the horsepower is developed and the same is true for oxy-boost. The very earliest oxy-boost installations in the glass industry were attempted with use of existing oxy-fuel burners - burners designed for use in the steel industry. These high velocity, low luminosity burners were totally inadequate for glass manufacturing, where heat transfer by radiation is the key to high melt rates and efficiencies.

The need for a high luminosity burner for the glass industry led to the development of the Cleanfire® Gen 1. burner. This tube-in-tube style burner utilized a retractable and tapered fuel nozzle such that the oxygen velocity could be matched with that of the fuel velocity to minimize mixing and thereby increase flame luminosity. The burner was connected to a refractory block, which not only provided a straightforward technique for mating the burner with the furnace superstructure, but also improved flame luminosity by promoting fuel decomposition or cracking within the burner's quarl; serving as a pre-combustor. While providing a huge step forward for the glass maker, the round and tube-in-tube nature of the burner produced a flame that was indiscriminate in the direction in which the radiation from the flame would be emitted, beyond that forced by temperature differentials within the melting chamber – see Figure 4 below.

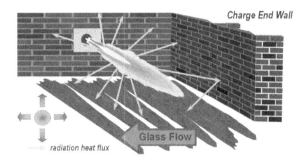

Figure 4 – Operation of a Tube-in-Tube Oxy-Fuel Burner

During this time, development engineers at Air Products were working on optimizing a burner for completely converting glass melters to oxy-fuel firing[1]. The Fourth Power Law of Radiation Heat Transfer suggested greatest benefit could be made of the high flame temperatures provided by an oxy-fuel flame. Design efforts also needed to focus upon the shape factor and the luminosity or emissivity of the flame (see Figure 6). It was found that not only did the adoption of a flat aspect ratio flame (flat flame) provide the greatest flame coverage above the glass melt, but also that the flat aspect ratio promoted higher levels of cracking of the fuel and hence higher flame luminosity[2].

A second and arguably the most beneficial enhancement on the tube-in-tube burner was the provision of under-shot oxygen beneath the main oxy/fuel flame. This under-shooting is accomplished by diverting a portion of the stoichiometric oxygen flow to a port in the burner block located beneath the primary fuel port. It was learned that injecting oxygen under the flame provides higher heat transfer rates to the glass and lower energy release to the superstructure refractory as evidenced by crown temperature decrease. It was also learned that staging the oxidant to delay mixing with the fuel could be used to produce lower levels of nitrous oxide emissions (NOx) generated within the sub-stoichiometric flame envelope.

Fourth Power Law of Radiation Heat Transfer

$$Q = \varepsilon \, A \, \sigma \, (T_{Flame}^4 - T_{Glass}^4)$$

Emmissivity   Shape Factor   Boltzmann Constant   Temperature Driving Force

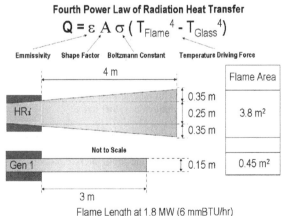

Figure 5 – Fourth Power Law Radiation Heat Transfer

The Cleanfire® HR™ burner combined both the flat aspect ratio and the oxygen under-shooting to produce a flame that is highly luminous, has high glass coverage, and is almost infinitely flexible via the use of an adjustable staging valve.

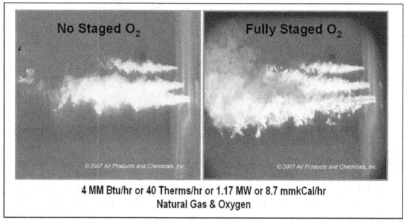

4 MM Btu/hr or 40 Therms/hr or 1.17 MW or 8.7 mmkCal/hr
Natural Gas & Oxygen

Figure 6 – Impact of Staging on the Cleanfire HR Oxy-Fuel Burner at a Fixed Firing Rate.

The key benefit to the under-shot oxygen staging was that now radiation flowed predominantly beneath the flame, directly towards the glass load. The sub-stoichiometric upper regions of the flame, provide lower NOx emissions and also produce an optically dense cloud which prevents heat from flowing towards the crown. These factors combined produce unequalled specific melt rates and fuel efficiencies.

However, the low momentum design of the Cleanfire HR burner produced a flame that was not well suited for the high-turbulence environment experienced when operating within an air-fuel fired furnace; i.e. in the role of an oxy-boost burner. Trials with other flat shape/high aspect ratio oxy-fuel burners produced very unstable flames. For example, when installed in the pre-port #1 cross-fired boost position in a float glass melting furnace, the flat aspect ratio oxy-fuel flame was seen to alternatively lap the adjacent superstructure or disappear under the charge endwall depending on the firing side of the air-fuel furnace. This instability could never be considered beneficial to refractory longevity!

Oxy-boost continued to be implemented with tube-in-tube burners, while operators of full-oxy-fuel furnaces enjoyed the benefits of a sculptured oxy-fuel flame.

The key benefit to the under-shot oxygen staging was that - with it - now radiation flowed predominantly beneath the flame, directly towards the glass load. In addition, the sub-stoichiometric upper regions of the flame provide lower NOx emissions and also produce an optically dense cloud which prevents heat from flowing towards the crown. These factors combine to produce unequalled specific melt rates and fuel efficiencies.

However, the low momentum design of the Cleanfire HR burner produced a flame that was not well suited for the high-turbulence environment experienced when operating within an air-fuel fired furnace; i.e. in the role of an oxy-fuel boosting burner. Trials with other flat shape/high aspect ratio oxy-fuel burners produced very unstable flames. For example, when installed in the pre-port #1 cross-fired boost position in a float glass melting furnace, the flat aspect ratio oxy-fuel flame was seen to alternatively lap the adjacent superstructure or disappear under the charge endwall depending on the firing side of the air-fuel furnace. This instability could never be considered acceptable for furnace operation and especially not beneficial to refractory longevity!

Therefore, oxy-fuel boosting systems continued to be implemented with tube-in-tube burners, while operators of full oxy-fuel furnaces enjoyed the benefits of a sculptured oxy-fuel flame made possible with the introduction of the Cleanfire® HRi™ burner.

The Advanced Boost Burner

To address this shortcoming in the industry, Air Products designed and launched the Cleanfire® HRi™ advanced boost burner. This burner, shown in figure 7 embodies all of the advances made with the oxy-fuel HR and HRi burners that were developed for full-furnace conversions, and is designed to operate in the turbulent atmosphere of the air-fuel furnace.

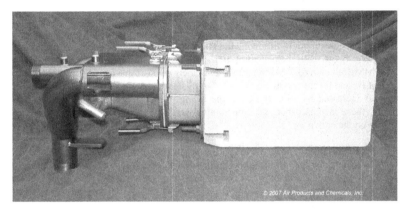

Figure 7 – The Cleanfire® HRi ™ Advanced Boost Burner

The design brief for the burner was to secure all the benefits of flat-flame combustion and to make it usable in an air-fuel environment. If accomplished, this would allow the technology to be used in oxy-fuel boosting applications – providing better results and with no downside as compared to the round shape tube-in-tube burners. Specific design goals included:

- To provide a stable and adjustable flame
- To provide discriminate heat transfer, preferentially towards the glass load
- To provide a low maintenance installation
- To provide lower pollution emissions than the current state-of-the-art
- Easy to operate

RESULTS

Within two years of market introduction, more than 40 Cleanfire HRi advanced boost burners have been installed in oxy-fuel boosting installations around the world. The following section details some of the results experienced with this new burner.

One of the early HRi burner installations was on a float furnace with regenerator problems affecting ports one and two. In this instance, 19 MMBtu/hr of natural gas was fired through the HRi oxy-fuel boosting burners. Optical temperature surveys showed that the boost burners increased refractory temperatures adjacent to the flame by only 36°F (20°C). This was testament to the directional heating ability of the HRi burners.

A furnace in Europe was operating oxy-fuel boosting with Gen 1 tube-in-tube burners to maintain a production rate of 700 mTPD. Following installation of the HRi burners, at the same production rate of 700 mTPD, the boost burners were turned down to 79% of the original heat input required with the tube-in-tube burners. The new burners initially shortened the batch blanket travel when installed at the same firing rate as the tube-in-tube burners. The reduction in fuel and oxygen amounts to a €280,000 ($400k) annual savings to operate this furnace.

Another European float furnace achieved a 45 mTPD production increase using an additional 2.56 MMBtu/hr (1682 kW) of fuel on the furnace (net air/fuel + Boost and compensating for 3% increase in cullet consumption) following the HRi oxy-fuel boosting installation. In this case, the benefits of oxy-boost really helped to optimize the furnace, bringing in the additional load at only 3.06 MMBtu/tonne (3.2 MJ/tonne).An extremely good result bearing in mind the energy requirement typical for the production of 45 TPD of float glass.. In this instance, furnace bottom temperatures increased by 19°C (66°F) at the filling pocket and 12°C (54°F) at the hot spot. Amazingly, these increased bottom temperatures were achieved with no increase in hot spot temperature on the crown.

One float line, again in Europe, achieved an 8% production increase, using only 2.3 MMBtu/hr (1500 kW) of additional fuel on the furnace. At this site, the customer achieved almost 3 times as much extra production per unit of oxygen used as they had achieved with their tube-in-tube oxy-fuel burners.

At a leading US float glass manufacturer who was using a common NOx reduction technique which tends to produce high levels of CO in the exhaust, the Air Products' Cleanfire HRi oxy-fuel boost system was used to reduce CO levels by greater than 60%. This eliminated the reducing environment that was causing rapid checker brick deterioration. In addition to the reduction of CO, the glass maker also reduced carryover by 20% and realized a furnace efficiency improvement of 6.5%. Their experience with this new technology was so compellingly positive that the glassmaker added the same system to another new furnace at another site.

At another furnace, this time in Asia, a glassmaker was operating with an oxy-fuel boosting system using tube-in-tube burners to achieve a production rate equal to 620 mTPD. While they had interest to trial new technology, they were not interested in a production rate increase. In this installation, the glassmaker wanted to confirm operational benefits prior to operating the oxy-fuel boosting system at lower firing rates with the new HRi burners installed. Following installation and return to stable operating conditions, the position of the batch blanket was confirmed to pull back significantly towards the charge endwall of the furnace. Since the glassmaker did not wish to change this visual process parameter that represents one key metric to optimization of their process, the fuel flow rate at the boosting burners was turned down to allow the batch blanket to return to its previous position for this production rate. This resulted in fuel savings similar to the other case studies referenced. Returning the position of the batch blanket to its desired position with the lower fuel flow rate also resulted in a 4°C (40°F) drop in crown temperature at this region of the furnace.

Several additional US glass producers have switched from other oxy-fuel boosting burners to the Cleanfire HRi burner technology – to date, mainly driven by the desire to improve furnace efficiency, reduce emissions, and to reduce charge endwall refractory temperatures. Typical results for this type of conversion include ~2% improvement in *net furnace efficiency* over installations already present but operating with other burners and 40-60°C (100-140°F) reduction of superstructure temperatures, and significant reductions in NOx and CO that are specific to the operation of each furnace installed.

CONCLUSION

In today's competitive world, we must get every ounce of productivity out of our assets. An oxy-fuel boosting system represents a relatively small investment that has proven the ability to significantly impact the operation of well-designed, very efficient and healthy cross-fired regenerative furnaces. The application can be used to increase production, decrease fuel consumption, decrease emissions, improve glass quality, and to extend the useful life of a furnace.

The trick to maximizing the benefit of the oxy-fuel boosting application is to utilize the fuel and oxygen as efficiently as possible. This is accomplished best with the technology provided in Air Products' Cleanfire® HR*i*™ advanced oxy-fuel boost burner which produces a flat, stable flame with directional heating to the glass load in the presence of an air-fuel combustion environment. With use of this technology, the boosting application provides net economic benefits so compelling that glassmakers are now making a commitment to use the boosting application through-out their entire furnace campaign as this is recognized as the lowest cost way to run their furnaces.

REFERENCES

1. Horan, W J, Slavejkov, A G and Chang, L, 'Heat transfer optimisation in a TV glass furnace', Proceedings of the 56th Glass Problems Conference, 1995.

2. Eleazer P B, Pastore S, Schemberg S, 'Burner Development Revolutionises Oxy-Fuel Firing', Glass International, March edition, , pp. 11-12 1996.

OXYGEN ENRICHMENT: RECOGNIZING AND ADDRESSING THE HAZARDS

Robert L. Martrich, Joseph W. Slusser, and Kevin A. Lievre
Air Products
Allentown, PA

ABSTRACT

Oxygen is the component of air that supports life. The very properties that make oxygen necessary to support life can make oxygen very hazardous. It is a colorless, odorless, tasteless, non-irritating oxidizing gas that comprises approximately 21% of our atmosphere. No physical properties permit the human body to detect its presence. You cannot see it, smell it, or taste it – there are no warning properties. In liquid form, it exists at extremely low temperature. When oxygen concentration is increased above the percentage found in the atmosphere, potential hazards exist. Oxygen enrichment occurs when the oxygen concentration in an atmosphere reaches a level at which the oxygen starts to increase its influence on other materials above that which would occur in ambient air. In other words the reactivity of the oxygen increases with its concentration and pressure. An example is the tremendous effect that increasing levels of oxygen has upon influencing the fire chemistry of combustible materials. This increased reactivity can lead to uncontrolled fires and explosions.

The use of industrial grade oxygen for combustion applications in the glass industry has increased dramatically over the past decade. The information presented is intended to help glass industry personnel that handle oxygen to have a greater awareness and assist them with performing their jobs safely. The paper reviews properties of oxygen, potential hazards, some aspects of system design, cleanliness requirements, personal protective equipment, and first aid procedures. The focus on hazards includes those associated with pressure, changes in fire chemistry, fog clouds, and cryogenic temperatures. These hazards need to be understood and considered with all actions in order to use oxygen properly. It is important that we recognize the hazards of enriched atmospheres and treat them with the same respect we have for pure oxygen.

Oxygen safety in glass manufacturing facilities is an important topic for review for many reasons. One reason is the lack of familiarity with the product. Due to the ever present need to improve operations - reducing fuel consumption, increasing productions rates, increasing glass quality, and reducing emissions – new uses of industrial grade oxygen have been implemented in melting operations at many glass plants. Emissions requirements have recently become a strong economic driver for this acceptance in use. Another reason is too much famiarity. At most sites, oxygen has been in use for decades in welding and other operations. Here, complacency becomes the issue that can lead to preventable injuries.

OXYGEN

Oxygen is an element that is found all around us in our everyday lives. It is a reactive element that is found in water, in most rocks and minerals, and in numerous organic compounds. It is a diatomic gas constituting 21 percent of our atmosphere and is capable of combining with all elements except the inert gases. It is active in physiological processes, and is a key component required for combustion. So, why is there any concern about the handling and use of oxygen?

Most people who use pure oxygen are aware of the hazards of and special precautions for handling this material. Unfortunately, fewer are aware that these hazards extend to oxygen mixtures. Therefore, it is helpful to review classification of the various mixtures of oxygen, the hazards of pure oxygen and oxygen mixtures, and how to safely handle these products. Most of the discussion that follows will concentrate on mixtures above 23% oxygen and pure oxygen, because operationally they should be treated with the same precautions and requirements.

OXYGEN MIXTURE CLASSIFICATIONS

The following information is based upon U.S. Compressed Gas Association (CGA) and ASTM International documentation, but the principles are repeated in other documents from the European Industrial Gas Association (EIGA) and the International Standards Organization (ISO).

The CGA defines oxygen-enriched mixtures or atmospheres as any mixture or atmosphere containing greater than 23% oxygen, since above this concentration, the reactivity of oxygen significantly increases the risk of ignition and fire. Materials that may not burn in normal air may burn vigorously in an oxygen-rich environment. Sparks normally regarded as harmless may cause fires. And materials that burn in normal air may burn with a much hotter flame and propagate at a much greater speed.

The CGA breaks oxygen and its mixtures into four distinct classifications:
Mixtures containing less than 5% oxygen in an inert: These mixtures are packaged and labeled in the same manner as inert gases like nitrogen and argon. For packaged gases, they use the same 580 valve connections, and are labeled with a nonflammable gas hazard diamond. These mixtures are considered inert.

Mixtures in the range 5 to 23% oxygen: These mixtures carry the same nonflammable gas hazard diamond as inerts, but are packaged with a 590 valve connection. This is intended to differentiate these mixtures from the inerts as they will support combustion but not enhance the danger of ignition beyond that of normal air.

Mixtures containing greater than 23% oxygen: These mixtures must be handled with all the precautions and care of pure oxygen. Mixtures at these levels start to change fire chemistry and enhance combustion similar to oxygen. They carry an oxidizer hazard diamond in addition to the nonflammable gas diamond and packaged gases are equipped with a 296 valve connection to distinguish them from the less reactive mixtures of oxygen. These mixtures may be ppm level components in otherwise pure oxygen, for example, 8 ppm methane in oxygen.

PURE OXYGEN

In the packaged gases, this material has its own dedicated 540 valve connection and an oxidizer hazard diamond.

There are other valve connection standards in use that may or may not distinguish between the various concentrations of oxygen. Therefore, labels must be relied upon to identify these mixtures because they may not be protected from interconnection.

HAZARDS OF OXYGEN AND MIXTURES

One of the challenges in educating people about the dangers associated with oxygen is their perception of the product. When most people think about oxygen, medical applications come to mind. Doctors, dentists, nurses, paramedics and first aid teams administer oxygen to people when they are sick or injured. These professionals would not allow us to intentionally inhale a hazardous material. This perception can downplay or at least minimize the respect for the hazards of oxygen. But let's look at oxygen from a different perspective.

Oxygen and its mixtures are packaged as compressed gases. They are supplied at pressures up to 4500 psig (300 bar). Pressure is stored energy, and if this pressure is released in an uncontrolled manner the resulting energy release can cause great damage or injury. This could occur if a cylinder is being handled without a protective valve cap and dropped. If the valve is struck hard enough, it could shear and the escaping pressure could propel the cylinder like a rocket. For more information on the proper handling of compressed gas cylinders refer to Air Product's Safetygram-10 "Handling, Storage and Use of Compressed Gas Cylinders". For more information on the consequence of shearing a cylinder valve, refer to Air Product's Safetygram-14 "Don't Turn a Cylinder Into a Rocket". All compressed gas cylinders demand proper handling and respect because of the tremendous energy stored within them.

Oxidizers carry additional hazards. To fully understand these hazards we must first understand some of the terms used:

- Adiabatic Heat: A process in which there is no gain of heat to or from the system. With respect to this document, it is the heat picked up by the gaseous oxygen from the rapid pressurization of a system.
- Autoignition Temperature: The lowest temperature required to ignite or cause self-sustained combustion in the presence of air and in the absence of a flame or spark.
- Flammable Range: The range of concentration in volume percent of flammable gas or vapor between the upper and lower flammability limits.
- Kindling Chain: The promotion of ignition from materials of low ignition temperatures to materials of higher ignition temperatures.
- Limiting Oxygen Concentration: The minimum oxygen concentration in volume percent (at a given temperature) in a gaseous mixture containing a fuel below, which a flame will not propagate.
- Lower Flammability Limit: The minimum fuel mixture in volume percent with air through which a flame will just propagate.
- Upper Flammability Limit: The maximum fuel mixture in volume percent with air at which a flame will just propagate.

When oxygen concentrations exceed 23%, oxygen enrichment begins and fire chemistry starts to change. Materials become easier to ignite because their flammable ranges start to expand and their autoignition temperatures begin to drop. This includes the materials of construction used in oxidizer systems, such as metals. This reactivity continues to increase not only with the concentration of oxygen, but also with pressure and/or temperature. In other words, oxygen contacting a material at 2000 psig is more likely to react with the material than at atmospheric pressure. In the case of a contaminant in a system, the contaminant may react and generate enough heat to start another material reacting. This is called the kindling chain. When temperature increases it can lower the amount of energy required to initiate a reaction.

Let's look at the basic fire triangle.

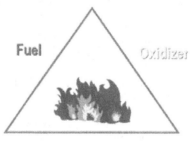

Fuel   Oxidizer

Source of Ignition

All three legs of the triangle must be present to produce a fire—a fuel, an oxidizer, and an ignition source. If asked to name some fuels, materials like wood, coal, oil and gas would be mentioned. But would anyone list materials like aluminum, steel, stainless steel? What is the primary reason we can light a piece of wood with a match but not a steel rod? The ignition temperature of the wood is much lower than that of the steel rod and the heat from the match is sufficient for ignition. Remember what we said about fire chemistry and oxygen—as the oxygen concentration increases, the autoignition temperature decreases. So materials that cannot be ignited in normal air may burn readily in oxygen enriched atmospheres. With this in mind, it is easy to see that in an oxidizer system we have two legs of the fire triangle present. All that is required for an ignition is an energy source.

Now let's consider ignition sources. Typical sources of ignition would be fire, open flames, sparks, cigarettes, etc. But that is in the world of normal air, not oxygen-enriched atmospheres. Remember the definition of autoignition temperature—the lowest temperature required to ignite a material in the absence of a flame or spark. Could gas velocity, friction, adiabatic heat or contamination provide ignition sources? Yes.

In the case of gas velocity, it is not the flow of gas that can cause ignition, but a particle that has been propelled by the gas and impacts the system with sufficient force to ignite. The heat generated may be sufficient to start a fire depending on the material impacted. Friction from a component malfunctioning or operating poorly can generate heat. Friction between two materials generates fine particles, which may ignite from the heat generated.

Adiabatic heat is sometimes confused with the heat of compression. The heat of compression causes the temperature of a system to rise. An example would be a tire pump. The barrel or compression chamber builds heat as the pump compresses air. This process occurs relatively slowly and the system takes on the heat. Adiabatic heat is caused by the rapid pressurization of a system where the gas absorbs the energy and the gas temperature rises. This heating occurs at the point of compression or the point where the flow of gas is stopped, such as at a valve or regulator seat. Depending on the material in use where the hot gas impinges, the heat may be sufficient to ignite the material.

All of these energy sources can be enhanced by the presence of a contaminant. Contaminants are typically easier to ignite than the components of the system. If they react with the oxygen, they may generate sufficient heat to propagate a reaction to the system. Or as in the case of the pipe in Figure 1, they may react so strongly as to compromise the system.

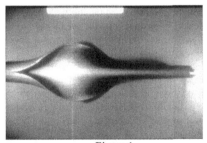

Figure 1

Figure 1 depicts a section of an oxygen pipeline that ruptured. Here's how this happened. The oxygen supply line at an installation needed to be extended an additional 150 feet (45.7 meters). The line was degreased, but shop air was used to clean the line rather than clean, dry nitrogen. Most shop air is compressed with a hydrocarbon compressor, and in this case, the compressor did not have a clean-up system to remove any trapped oil. When the line was purged, a thin film of hydrocarbon oil coated the interior of the pipe. The pipe was put into service and the operation went smoothly until at the end of the shift when a valve at the downstream end was closed. This stopped the flow and the oxygen heated as it compressed against the valve seat. The compression provided enough energy to react with the hydrocarbon oil and a deflagration occurred. The speed of the pressure wave was such that it duplicated the line rupture, as depicted in Figure 1, every 15 feet (4.57 meters). This is a very good example of how materials that burn in normal air can react in oxygen.

It is critical to keep non-typical ignition sources in mind when designing systems for oxygen use. Some applications are very vulnerable to ignition. For example, the elastomers used in valves and regulator seats have lower ignition energies than metals. Since particle impingement or adiabatic heat can be directed at these valves and regulators, they are particularly susceptible to ignition.

Let's take a look at a carbon steel pipeline used to provide oxygen to a customer. Because most of these pipelines have large diameters, economy and availability make carbon steel the material of choice. Carbon steel is an excellent fuel when used in oxygen service. In fact, due to the operating pressure, the pipelines contain a flammable mixture. In order to prevent fires in these pipelines, ignition sources must be considered. Since these lines are underground, external sources are usually not a problem. However, carbon steel is prone to rust which can generate particles. If a particle is picked up in the flow of oxygen, the particle may impinge on part of the system. If the impingement generates sufficient heat, it may provide a source of ignition energy. Carbon steel pipelines in oxygen service must be designed and operated to minimize the potential hazard. This is accomplished by following the gas velocity limitations set forth in CGA Pamphlet G-4.4. By limiting the velocity, the particle impact energy is reduced which minimizes the chance of an ignition.

The above examples show how contamination in a system can enhance the potential for a reaction. It must be stressed that systems must not only be cleaned to oxidizer service requirements on initial construction but must be maintained in that condition of cleanliness. Contamination of materials like hydrocarbons or contaminants that may be in the form of particles can initiate a system fire.

These are potential problems in oxidizer systems, but what happens when oxygen enters the atmosphere? If oxygen leaks or escapes into the air as part of a normal process, oxygen enrichment may occur if the ventilation is insufficient. If the surrounding air becomes oxygen enriched, the same changes in fire chemistry take place. In other words, materials become easier to ignite because their autoignition temperatures begin to decrease and their flammability ranges increase. Remember materials that do not ordinarily burn in air may ignite, and materials that do burn in air will burn hotter and faster. Most areas where enrichment is known to be a possibility are posted with signs warning about smoking or open flames. The real danger is if oxygen enrichment occurs and a person carries the hazard with them when leaving the area.

Clothes can trap and hold the oxygen-enriched atmosphere in the countless air spaces within the weave of the fabric. Hair poses the same problem. If someone comes in contact with an enriched atmosphere, it is extremely important to isolate them from any source of ignition for at least 30 minutes. Get them to fresh air, pat their clothing, and run fingers through or wet their hair. Think about how easily cloth or hair burns in air and then try to imagine how they would burn in an oxygen-enriched environment.

Figure 2   Figure 3

Figure 2 shows a normal cotton work shirt, stuffed with paper and saturated with oxygen. Inside the shirt is a flash bulb with the glass removed exposing the magnesium wool. This bulb is connected to a battery to provide ignition. The results of the ignition are shown in Figure 3. People must be aware of the hazards of oxygen enrichment, since oxygen is not detectable by by human senses. It is colorless, odorless, tasteless and non-irritating. Likewise, people must be aware where oxygen enrichment is possible, such as at a vent, a leak, or failure of a ventilation system. Sometimes people intentionally saturate themselves with oxygen without recognizing the danger. Think about the person using who may use the flow of oxygen from an oxy-acetylene torch to blow off dirt or to cool down.

SAFE HANDLING OF OXYGEN AND MIXTURES

It is important that all personnel handling oxygen and its mixtures be aware of its hazards and trained in the use of proper procedures and personal protective equipment. The MSDS is the primary source of hazard data. All personnel handling a material should be thoroughly familiar with the MSDS, including personal protective equipment recommendations.

As mentioned earlier, oxygen mixtures are classified, labeled, and valved differently. The primary method of product identification is the label. Mixtures with oxidizing potential will have a yellow oxidizer hazard class diamond in addition to the green nonflammable gas diamond. Do not let the presence of the green nonflammable label lessen the warning of the oxidizer hazard.

The valve connections provided may prevent the possible connection of an incompatible material, but must not be relied upon for identification of the mix. If the mixture label and valve connection do not comply with the above-mentioned classifications, DO NOT use the cylinder until consulting with the supplier. NEVER use adapters or change connections to accommodate the connection of a cylinder to a system. If a cylinder does not connect easily, do not force, put the cylinder aside, label as to problem and contact the supplier.

From an operating perspective, areas of possible oxygen enrichment must be identified and plans implemented to address these potential hazards. These plans should consider safety systems including but not limited to training, signage, monitoring and ventilation. Systems using mixtures of oxygen 5% and greater at high pressure (greater than 450 psig / 30 bar) should be cleaned for oxygen service. Most compressed air systems use compressors that are hydrocarbon lubricated. However, the pressures at which they operate do not pose reaction problems and cleaning these systems is not practical because they are self contaminating from the compressor oil. However, materials of construction and system design requirements for enriched oxygen concentrations must go beyond cleaning.

Plant systems must be designed with the special considerations required for the safe handling of oxygen. There are several documents available to help design systems and equipment for the safe handling of oxygen. They include but are not limited to Compressed Gas Association Pamphlet G-4.4, "Industrial Practices for Gaseous Oxygen Transmission and Distribution Piping Systems", and the ASTM International Standard G 88, "Designing Systems for Oxygen Service". The European Industrial Gas Association and the International Standards Organizations also publish related documents including IGC Doc 13/02/E, "Oxygen Pipeline Systems"..

Designing and building these systems requires an intimate knowledge of oxygen and how it reacts with the materials it contacts. Basic design considerations include but are not limited to control and avoidance of unnecessarily high temperatures and pressures; cleanliness; elimination of particles; minimization of heat of compression; avoidance of friction and galling; minimization of resonance with direct flow paths; use of hardware that has a proven history in oxygen service; minimizing available fuel and oxygen through materials selection and system volume; anticipation of indirect oxygen exposure from system failures; and design of systems to manage fires using techniques, such as, fire stops, and automatic extinguisher systems. The first step in constructing any system for oxygen should be to consult your supplier.

CHROMIUM VI: CONCERNS, COMPLIANCE AND CONTROLS

Terry L. Berg
CertainTeed Corporation
Blue Bell, PA

Carlos E. Davis
CertainTeed Corporation
Kansas City, KS

ABSTRACT

It has been determined that airborne particles of hexavalent Chromium, Cr^{+6}, pose a number of possible health threats, of which, both employers and employees need to be made aware. Understanding and implementing OSHA STANDARD 29CFR-1910.1026 and its supporting documentation will provide protection for your employees, contractors and the community-at-large. This paper is presented by permission of Saint Gobain Corporation, Valley Forge, PA.

CHROMIUM VI: CONCERNS, COMPLIANCE AND CONTROLS

Chromium

Chromium is a naturally occurring element found in rocks, animals, plants, soil, and in volcanic emissions. Chromium (Cr) has three common valence states: (0), (+3), and (+6). The first is most commonly used in the manufacturing of steel and other alloys. Cr^{+3} is an essential nutrient which promotes the action of insulin so that sugars and fats can be used by the body. In its mineral form, Chromite, Cr^{+3} is used in the making of refractory bricks and blocks for high temperature industrial applications. Cr^{+3} and Cr^{+6} are also prepared in various chemical compounds for use in chrome plating, leather tanning, dyes and pesticides used to protect wood. Cr^{+6} is also a by-product of certain industrial reactions such as welding operations and by exposing Chromium refractories to high temperatures over time. With the myriad uses of Chromium and approximately 300,000 U.S. workers exposed to Chromium compounds in the workplace[1], it becomes very important to understand the long term effects of such exposure. Here is where we enter the field of epidemiology.

Epidemiology

Simply stated, epidemiology is the study of the distribution and causes of disease in populations. More formally stated:

"Epidemiology is the study of factors affecting the health and illness of populations, and serves as the foundation and logic for interventions made in the interest of public health and preventive medicine."[2]

I think at this point it would be interesting to look briefly at some people and historic events which have provided the foundation for this field of study.

The History of Epidemiology

Hippocrates is considered by many to be the "father of epidemiology."[3] It is reported that he kept a collection of notes on diseases common to his day and the cities which were most affected by them. He gave us two terms that we still use today:

- **endemic** – diseases found in some places but not others and
- **epidemic** – disease found at some times but not others.

Girolamo Fracastoro, in 1546, wrote a book entitled _De Res Contagiosa_ or "The Seeds of Contagion"[3]. He had observed that people who came in contact with sick people often contracted the disease themselves. He felt that this contact allowed these "invisible" seeds to pass from one person to another. He immediately fell into disfavor with the Roman Catholic church which taught that demonic possession was the cause of illness and with the "scientists" of the day who held to the Galenic theory of "miasmas" (vapors from swamps or dead bodies) being the cause of all diseases. It took over 300 years before the "germ (seed)" theory started to gain acceptance.

The invention of the microscope in 1674 by Anton van Leeuwenhoek still wasn't enough to break the strongholds of the prevailing medical opinions of the day. It is hard to imagine when they looked at pond water through this marvelous invention and saw for the first time the abundance of formerly invisible living creatures swimming about that they could still cling so strongly to their wrong beliefs. The world still wasn't ready for the germ theory.

In 1847, Dr. Ignaz Simmelweis at the Vienna Hospital, noticed a huge disparity in the mortality rates between the two maternity wards there. In the first ward, attended to by medical students the rate was 122 deaths per thousand mostly due to infection, while the second ward, attended to by mid-wives the rate was only 33 per thousand. He took the time to observe the activities of both groups to see if he could determine any correlation. The mid-wives spent their time knitting, rocking babies and chatting between deliveries but the medical students were busy working on cadavers or making rounds with their instructors. Simmelweis started asking that all of the medical students wash their hands before going into the delivery ward. When the mortality rate dropped to 12 deaths per thousand over the next few months, Simmelweis began promoting his own version of the "germ" theory, for which he was promptly suspended for 6 months.[3]

In 1854, there was a cholera outbreak in the Soho district of London. Dr. John Snow has provided for us one of the earliest attempts at charting a disease by location (endemic study). His chart (public domain) appears below:[4]

He made a plot of the Soho District and marked the location of the public water pumps along with the deaths due to cholera. He observed a huge cluster around the pump on Broad Street and so he removed the handle from the pump, forcing the people to go to other wells. The number of deaths began to decline. It can be argued, of course, that the epidemic was already in decline or that the people, who had already been exposed before the handle was removed, weren't going to get the disease anyways. It was still a daring idea in an age of medical ignorance.

Louis Pasteur, (1827-1912), did extensive research not only in tracing diseases back to their source but also develop the means to deal with them. The pasteurization process, although quite probably his best known contribution, may be less significant than his pioneering work in the area of bacterial vaccination. He even developed a vaccine for rabies through his studies. Through his work, the "Germ Theory" finally gained dominance in the medical community.[3] and lead to the development of what is known as the "Epidemiology Triangle" between environment, agent and host.

Robert Koch, wrote a treatise in 1890 based on Pasteur's discoveries which formalized the basic rules of epidemiology entitled, *The 4 Postulates of Cause and Effect.*[3] The four postulates are thus stated:

1. *If a parasite is present in all cases, at all stages, and*
2. *If it is not present in other diseases, and*
3. *If it can be isolated from the body in pure cultures, and*
4. *If that culture can cause the disease anew, then a cause and effect relationship exists.*

The postulates, as recently as 1965, have been revisited by Austin Bradford Hill, who further detailed criteria for assessing evidence of causation. These guidelines are sometimes referred to as the *Bradford-Hill criteria.* These are simply stated as:

1. Strength: A small association does not mean that there is not a causal effect.

2. Consistency: Consistent findings observed by different persons in different places with different samples strengthens the likelihood of an effect.
3. Specificity: Causation is likely if a very specific population at a specific site and disease with no other likely explanation.
4. Temporality: The effect has to occur after the cause (and if there is an expected delay between the cause and expected effect, then the effect must occur after that delay).
5. Biological gradient: Greater exposure should generally lead to greater incidence of the effect.
6. Plausibility: A plausible mechanism between cause and effect is helpful
7. Coherence: Coherence between epidemiological and laboratory findings increases the likelihood of an effect.
8. Experiment: "Occasionally it is possible to appeal to experimental evidence"
9. Analogy: The effect of similar factors may be considered.

Hill himself said "None of my nine viewpoints can bring indisputable evidence for or against the cause-and-effect hypothesis and none can be required sine qua non".[5]

Types of Studies

There are four basic types[6] of epidemiology studies: ecologic, case control, cohort and clinical trials. Ecologic studies look only at data about populations and geographic regions to try to develop associations. In case control studies, the history of patients with the same diagnosis and are strongly subject to bias of persons doing the investigation. The outcome is usually expressed as the **"OR"** (The odds ratio is the likelihood that those with the disease have been exposed to "X".) Cohort studies are used where a group is disease-free at the start and are observed over time as they are exposed. These outcomes are called the **"RR"** (The risk ratio or "relative risk" is the likelihood that those who were exposed to "X" will develop the disease.) Last is the clinical trial which on the surface would appear to be the most objective. Typically, a group of people with the same diagnosis is split into a test group and a control group. The test group gets the drug while the control group gets a placebo. The problem is multifold as drug companies fund most of the research and subjects are sometimes given inappropriate doses of the competitor's product or undesired results don't get reported at all.

What is the end result of all these studies in the courts? In United States law, epidemiology alone cannot prove that a causal association does not exist in general. Conversely, it can be (and is in some circumstances) taken by US courts, in an individual case, to justify an inference that a causal association does exist, based upon a balance of probability. Strictly speaking, epidemiology can only go to prove that an agent could have caused but not that, in any particular case, it did cause: "Epidemiology is concerned with the incidence of disease in populations and does not address the question of the cause of an individual's disease. This question, sometimes referred to as specific causation, is beyond the domain of the science of epidemiology. Epidemiology has its limits at the point where an inference is made that the relationship between an agent and a disease is causal (general causation) and where the magnitude of excess risk attributed to the agent has been determined; that is, epidemiology addresses whether an agent can cause a disease, not whether an agent did cause a specific plaintiff's disease."[7]

What all this means is that correlation doesn't always equal causation so therefore epidemiologists use the word "inferred" a lot. They realize that most outcomes are caused by a web consisting of many factors, such as, age, sex, fitness, diet, genetic make-up and emotional state and that statistics are not science. Someone once said, "People use statistics when they don't have any hard evidence."

Health Risks

In the case of Chromium (VI), Cr^{+6}, enough data has been collected and subjects studied, since 1912 in fact, that it appears that there are significant health risks. Workers who breath airborne Cr^{+6} on their jobs for many years may be at increased risk of developing lung cancer. Furthermore, breathing high level of Cr^{+6} can irritate or damage the nose, throat and lungs and cause lesions. Even simple contact with Cr^{+6} dust for prolonged periods of time can cause irritation and damage to the eyes and skin. For these reasons, every reasonable precaution must be taken to minimize exposure to or contact with Cr^{+6}.

OSHA STANDARD 29CFR 1910-1026

Because each work site is unique, it is impossible to provide a single model for an exposure - reduction program. The site project managers and subcontractors must work closely to jointly develop a program that is adapted to their site.

Several substances which can be released during furnace renovation & maintenance are classified as carcinogenic, such as:
• Crystalline silica classified by the International Agency (IARC) as carcinogenic to human (category 1)
• Chromium VI classified in several countries as carcinogenic to human (category 1)
• Refractory Ceramic fibres (RCF) classified as probably carcinogenic (category 2) by the European Union.
• Heavy metals including As, Ni, Pb

In most of the countries, there are specific requirements for the use of CMR (Carcinogenic, Mutagenic or toxic for Reproduction) substances such as the European directive 2004/37/EC in Europe. All are based on the same approach:
• Determination and assessment of risks
• Reduction and replacement
• Prevention and reduction of exposure

In order to recognize the presence of hazardous materials, the nature of all refractory materials to be removed or installed must be identified, along with any potentially associated contaminants coming from and/or formed in association with the glass or fumes with which they have been in contact.

It is important to know the history of a material before dismantling it as some hazardous substances are sometimes generated by transformation or contamination during the furnace lifetime. For example (non exhaustive):
• RCF & other fibres can partially convert to crystalline silica (Cristoballite) if used at temperatures in excess of 900 °C over extended periods.
• In case of Alcalis and Oxygen presence, and at a certain temperature, Chromium III can be oxidized into Chromium VI during the furnace lifetime.
• Refractories of furnaces lining are sometimes contaminated with arsenic, chromium, lead, selenium, …

In planning for activities involving potential worker exposure to hazardous substances, drawings of the existing installations or those to be built should be supplemented with an inventory of dust-producing operations. This inventory provides the basis for all subsequent preventive measures.

For each dust-producing operation, the following points should be addressed:
• nature of the operation and its duration;
• number of workers involved;
• nature of the hazardous materials to be removed or installed such as fibrous insulation and refractory materials, taking into account temperatures and duration of use;
• work methods to be used;
• handling and disposal of waste;
• possible presence of other contaminants;
• collective and individual protection measures required.

The work site (furnace, forehearth, waste chute and bin) should be isolated as much as possible. For example, an enclosure of plastic or canvas sheeting (or rigid plastic panels) and should be clearly demarcated. A gateway (vestibule) should be provided for operators going in and out the working zone. Another one might be necessary for transit of equipment, waste or other items....

Large sites are difficult to enclose, such as big float glass furnaces. In these cases, attention should be focused on isolating or enclosing the dustiest operations.

This preventive measure has two objectives:
a) to limit the dispersion of dust and
b) to clearly identify the work site.

Adequate ventilation and a negative pressure should be maintained in the working zone (for example a renewal of the air volume 6 times per hour can be recommended in case of presence of chromium VI) . This ventilation is also helpful to improve the comfort of operators (in case of furnace not completely cooled down, or hot ambient weather)

Furnaces, forehearths and their associated installation should be dismantled rather than demolished. Dismantling work should be carried out in the following sequence:

1. preliminary cleaning,
2. Removal of refractory materials
   • specific removal of RCF insulation,
   • dismantling and sorting of hazardous refractory materials.
   • dismantling and sorting of other refractory materials

Please note that dismantling is the most hazardous operation mentioned in this document, because of the following issues: operator's safety, materials handling and toxic agents generation such as crystalline silica, chromium VI, RCF or heavy metals, which should be minimized as much as possible.

During the preliminary cleaning, to prevent re-suspension of settled dust, all dusty surfaces in the work site should be cleaned first (e.g., furnace regenerator, furnace tank and roof, immediate surroundings, metallic structures).

The cleaning should be done using either a vacuum cleaner equipped with a high-efficiency filter (HEPA - High Efficiency Particulate Air Filters) or wet cleaning methods. Compressed air or use of domestic-type vacuum cleaners should be formally prohibited, along with dry sweeping.

After wetting, products should be removed as gently as possible and immediately bagged. If wetting is not possible, a local ventilation system should be used to capture dust as close as possible from the sources.

The refractory materials contaminated by chromates should be handled with precaution (adapted PPE should be worn), and be stored in a strong & air-tight (sealed) container such as double skin big-bags or similar for secondary treatment . As much as possible, the blocks should be loaded directly into the final container on-site (one-step operation), without temporary storage inside or outside on the ground

The last step of the dismantling operation is , of course, a thorough cleaning of the working zone and a careful removal of the enclosure.

Any exposure - reduction program will fail if workers do not or cannot follow it.

Sufficient and appropriate information and training can ensure program success. The employer must ensure that every worker intervening on the site is fully informed of the potential health risks associated with the different materials to be removed or installed, and sufficiently and appropriately trained to work with the materials. Workers must be aware of the exposure - reduction program - in particular the work procedures that they will be expected to follow in order to minimize dust. They must also be trained in the correct use of protective equipment.

The content of this training must be recorded and signed by workers.

Access to the site should be strictly limited to those workers who have been informed and trained. Examples of documents to be read & signed by subcontractors are mentioned in the bibliography.

BIBLIOGRAPHY

1. ATSDR.CDC*, Public Health Statement: Chromium, September2000

2. Last JM (2001). "A dictionary of epidemiology", 4th edn, Oxford: Oxford University Press.

3. Morabia, Alfredo. ed. (2004) A History of Epidemiologic Methods and Concepts. Basel, Birkhauser Verlag. Part I.

4. "Epidemiology" www.wikipedia.com

5. Hill AB. (1965). The environment and disease: association or causation? *Proceedings of the Royal Society of Medicine*, 58, 295-300

6. Milloy, Steven J..(2001). Junk Science Judo. Cato Institute, 1000 Massachusetts Avenue, N.W., Washington, D.C. 20001

7. Hennekens C.H. and Buring, J.E. (1987) , Epidemiology in Medicine.™ Mayrent, S.L (Ed.), Lippincott, Williams and Wilkins

29 CFR-1910.1026, www.osha.gov   [Chromium (VI)]

29 CFR-1910.134, www.osha.gov   [Respiratory Protection]

29 CFR-1910.141, www.osha.gov   [Protective Clothing]

29 CFR-1926.51, www.osha.gov   [Wash Room Requirements]

8. Saint-Gobain Corporation, EHS Guidelines For Building, Renovation And Maintenance Of Glass-Melting Furnaces, 2006

*Agency for Toxic Substances and Disease Registry   www.astdr.cdc.gov

EXAMINATION OF A USED CHROME-ALUMINA MONOLITHIC LINING FROM AN INSULATION FIBERGLASS (C-GLASS) ELECTRIC MELTER

Howard Winkelbauer and Mathew Wheeler
North American Refractories Company
Cincinnati, Ohio

ABSTRACT

Cast and fired in place chrome-alumina monolithic furnace sidewalls have proven to be a viable alternative to using preformed and kiln-fired sidewall blocks reducing both furnace downtime and rebuilding costs. To date, there have been 14 cold top electric (CTE) melters and one gas/oxy furnace cast with monolithic glass contact sidewalls.

This paper discusses petrographic, apparent porosity, and chemical changes in core-drilled samples from a cast furnace lining. These changes confirmed that furnace life using cast sidewalls is comparable to that of fired sidewalls of similar composition. New evidence indicates that the originally reported cast furnace life of 4-5 years may be extended even further.

INTRODUCTION

Historically, monolithic refractories are used in the glass industry as repair materials to extend the life of operating furnaces. Monolithic refractories have long been the leading refractory used in many other industries, and are equivalent to or surpass the performance of ceramic bonded materials [1]. When a furnace rebuild was necessary, the glass contact zones are typically lined with pre-formed and kiln-fired shapes or fused cast refractories. For example, prior to 1980 most furnaces for melting insulation fiberglass were built primarily with fused cast alumina-zirconia-silica materials as sidewalls and top pavers and fused cast alumina-chrome materials were used in the throat and other critical areas of the furnace. Since then, ceramic bonded alumina-chrome products having 16-95% chromia have been widely used in insulation fiberglass furnace construction [2].

An alternative to this approach was presented at the 2002 Conference on Glass Problems in Columbus, OH [3]. This approach involved the use of cast 50% chrome-alumina monolithic linings. This effort was a joint project undertaken by Owens-Corning and North American Refractories Company. Since the beginning of this program, the sidewall refractories in 14 C-glass furnaces have been cast. The first complete CTE melter was cast in February, 1997. Twelve months later a second and then a third furnace were cast. This paper concerns the zonal analyses of two core-drilled plugs taken from the cast sidewall of the third furnace after it had been brought down for repairs to the forming machines and before it was recast with the 50% chrome-alumina monolithic and placed back into service.

The cast-in-place furnace lining that is the subject of this paper was in operation for 48.3 months, until April 2002, and had produced 220,100 tons of glass. It had been taken out of service due to repairs that were required to be made to the fiberglass production machines, not the refractory, providing an opportunity to inspect and obtain core samples of the sidewalls for analysis. The inspection of the interior of the furnace confirmed the furnace's operational history and the consensus from the parties involved in the inspection was that the cast lining still had significant service life remaining. The core drilled samples were retained for analysis and to hopefully provide insight into why the cast in place lining may work as well as or better than pre-fired block.

Figure 1 is a photo of the completed cast sidewall after the furnace had been cast back to its original 8" lining thickness with the 50% chrome-alumina castable. The original lining consisted of 30% alumina-chrome fired blocks that were in service for approximately 4 years. The furnace was cast back to its original 8" thickness with a wall that was sloped 4" from top to bottom to provide support for the bulk of the castable which was located in approximately the middle of the cast wall.

Figure 1. Corner of the furnace shown prior to start up in which the sidewalls were cast with a 50% alumina-chrome castable (JADECAST 50).

Figure 2 illustrates the sidewall construction and provides some detail as to the wear patterns of the original and cast linings. As can be seen in the figure, both the original 30% chrome and the cast sidewalls exhibited the typical "C" shape wear pattern that is found in a cold top electric melter. The 50% chrome-alumina cast walls had been eroded to a lesser extent than the original lining.

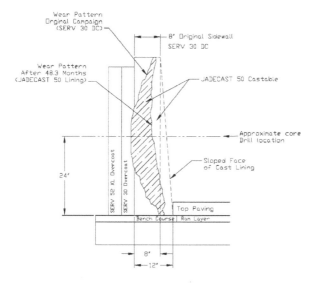

Figure 2. Section view showing schematic of original and cast lining.

Figure 3. Photo of the Used Lining

Figure 3 is a photo of the rear-wall of the subject furnace after it was drained. After minor repairs, this furnace was cast a second time and ran for about five years.

LEFT CORE-DRILLED SAMPLES

Core-drilled samples were taken from both the left and right side endwalls of the furnace before the furnace was recast and placed back into service. The left core was about 15" in remnant length and consisted of the following materials shown in Table 1:

Table 1
Sections in the Left Core

Depth from Hot Face, inches	Section	Brand	Composition / Application
0 to 3.5	L1, L2	JADECAST 50	50% Alumina-Chrome Castable (Cast Lining)
3.5 to 6.5	L3	SERV 30 DC	30% Alumina-Chrome Cast Fired Block (Original Sidewall Blocks)
6.5 to 10.75	L4, L5	SERV 30 with SHAMROCK 192 Patch	30% Alumina-Chrome Fired Brick with 22% Alumina-Chrome Patch (Overcoat and Patch for Bonding Overcoat)
10.75 to 15	L6, L7, L8	SERV 52 XL	50% Alumina-Chrome Fired Brick (2nd Overcoat Layer)

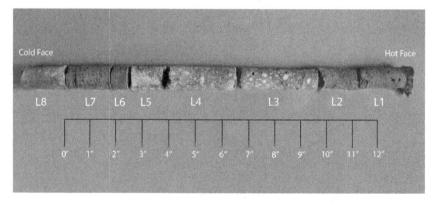

Figure 4. Photo showing the core-drilled sample taken from the left sidewall.

On the left side where the core drill sample was taken the original thickness of the JADECAST 50 was estimated to be 7". Sections were taken from each refractory quality for apparent porosity, chemical analysis and microscopic examination.

Figure 5. Porosity Changes in Left Core Drill

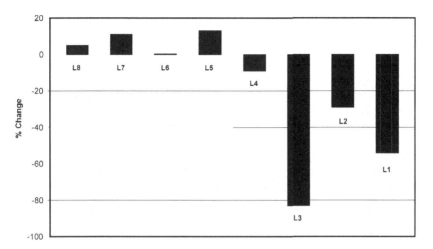

Figure 5 shows the magnitude of the apparent porosity changes in the left core drill. A decrease in apparent porosity was caused by glass infiltration into the sample. As expected, the greatest change occurs in the hot face section. Sample L3 shows a significant decrease in apparent porosity because it was the exposed hot face of the original fired block before the furnace was relined. The minor increases in apparent porosity in the deeper sections of the lining (L7 and L8) may have been caused by permanent expansion of the refractory from long-term exposure to elevated temperatures. This is occasionally seen in the unaltered sections of used refractories. The minor increase in apparent porosity of L5 may have been caused by some patch adhering to the brick sample.

Figure 6. Amounts of Infiltrated Oxides in Lining

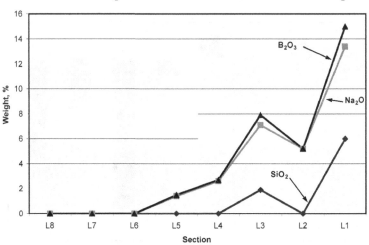

Figure 6 shows the amounts of infiltrated glass in the sections from the left core drilled sample. Both silica and boron oxide show a major drop in section L2 (2 to 3.5" depth from the hot face). Only soda and boron oxide had infiltrated into section L2. Likewise, L4 shows a similar drop in silica and soda from the amounts in the previous exposed hot face section (L3). Again, notice that soda remains at elevated levels until the beginning of the 50% alumina-chrome overcoat brick. Apparently, molten glass did not migrate into the last overcoat block.

Sections L 1 and L2 (JADECAST 50)

Figure 7   JADECAST 50 - Unused; 100x

1. Fused Alumina-Chrome Grain
2. Fused Chrome-Alumina Grain
3. Chromic Oxide Enriched, Cement
   Bonded Matrix
4. Void/Epoxy

Figure 7 shows the microtexture of unused JADECAST 50 at 100X. It is characterized by a combination of a coarser fused alumina-chrome grain (shifted Corundum x-ray diffraction pattern) and intermediate fused chrome-alumina grain (shifted Eskolaite pattern) in a chromic oxide enriched (Eskolaite), calcium aluminate cement bonded matrix.

Figure 8   JADECAST 50 - Used (L1); 100x

1. Fused Alumina-Chrome Grain
2. Fused Chrome-Alumina Grain
3. Glass Penetrated Matrix
4. Void/Epoxy

Figure 8 shows the altered microstructure of Section L1. It shows significant glass penetration throughout the sample. Some reaction along the edges of the fused grains was noted. The matrix primarily consisted of individual chromic oxide laths (Eskolaite) in an amorphous Na-K-Ca-Al-Si-O glass. Apparently, these fine laths form a three dimensional refractory structure that bonds to the coarser grains and helps the altered portion of the lining resist erosion.

Figure 9   JADECAST 50 - Used (L2); 100x

1. Fused Alumina-Chrome Grain
2. Fused Chrome-Alumina Grain
3. Glass Penetrated Matrix
4. Void/Epoxy

Figure 9 shows that even though silica and boron oxide from the infiltrated glass was absent in this section, significant reaction still occurred due to the presence of soda presumably from the infiltrated glass.

Section L3 (SERV 30 DC)

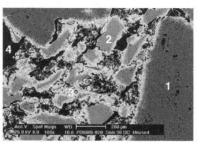

Figure 10   SERV 30 DC - Unused; 100x

1. Sintered Alumina
2. Intermediate Size Sintered Alumina Grain
3. Cement Bonded Matrix
4. Void/Epoxy

Figure 10 shows the microstructure of unused SERV 30 DC at 100X. It is characterized by a combination of sintered alumina (Corundum) and fine chromia (Eskolaite) in a calcium aluminate cement-bonded matrix.

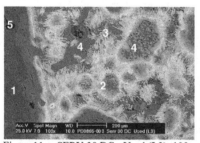

Figure 11   SERV 30 DC - Used (L3); 100x

1. Sintered Alumina Grain
2. Intermediate Size Sintered Alumina Grain
3. Glass Penetrated Matrix
4. Na-K-Ca-Al-Si-O Glass
5. Void/Epoxy

SEM examination of the left core drill sample (L3) showed glass penetration throughout the sample. Some reaction along the edges of the coarser sintered alumina (Corundum) grains as well as disruption of the intermediate size grains was noted. The matrix primarily consisted of individual chromic oxide laths (Eskolaite) in an amorphous Na-K-Ca-Al-Si-O glass.

Sections L4 and L5 (SERV 30)

The microtexture of unused SERV 30 was characterized by a combination of sintered alumina (Corundum) and fine chromia (Eskolaite) with an aluminum orthophosphate ($AlPO_4$) bond (Figure 6).

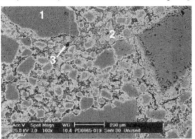

Figure 12   SERV 30 - Unused; 100x

1. Sintered Alumina Grain
2. Phosphate Bonded Matrix
3. Void/Epoxy

SEM examination of the SERV 30 from the left core drilled sample (L4 and L5) showed minor sintering and limited penetration by an amorphous Na-K-Ca-Al-Si-O glass into the pore structure; a significant portion of the pore structure remained unpenetrated (Figures 13 and 14).

Figure 13   SERV 30 - Used (L4); 100x

1. Sintered Alumina Grain
2. Intermediate Size Sintered Alumina Grain
3. Matrix w/Glass Penetration
4. Void/Epoxy

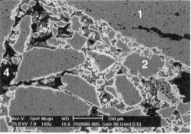

Figure 14   SERV 30 - Used (L5); 100x

1. Sintered Alumina Grain
2. Intermediate Size Sintered Alumina Grain
3. Matrix w/Glass Penetration
4. Void/Epoxy

Sections L6, L7 and L8 (SERV 52 XL)

The microtexture of unused SERV 52 XL was characterized by a combination of fused alumina-chrome (shifted Corundum pattern) and some fused chrome-alumina (shifted Eskolaite pattern) grains in a high chrome (Eskolaite) matrix (Figure 15).

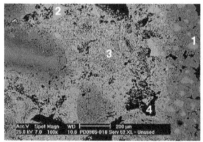

Figure 15   SERV 52 XL - Unused; 100x

1. Fused Alumina-Chrome Grain
2. Fused Chrome-Alumina Grain
3. High Chrome Matrix
4. Void/Epoxy

SEM examination of the used SERV 52 XL from the left core-drilled sample (L7 and L8) showed no alteration (Figures 16 and 17).

Figure 16   SERV 52 XL - Used (L7); 100x

1. Fused Alumina-Chrome Grain
2. Fused Chrome-Alumina Grain
3. High Chrome Matrix
4. Void/Epoxy

Figure 17   SERV 52 XL - Used (L8); 100x

1. Fused Alumina-Chrome Grain
2. Fused Chrome-Alumina Grain
3. High Chrome Matrix
4. Void/Epoxy

RIGHT CORE-DRILLED SAMPLES

Table 2
Sections in the Right Core

Depth from Hot Face, inches	Section	Brand	Composition/Application
0 to 5.8	R1, R2, R3	JADECAST 50	50% Alumina-Chrome Castable (Cast Lining)
5.8 to 7.8	R4	SERV 30	30% Alumina-Chrome Cast Fired Brick (Overcoat Block)
7.8 to 9.8	R5, R6	SERV 30 + SHAMROCK 192 Patch	30% Alumina-Chrome Fired Brick with 22% Alumina-Chrome Patch (Overcoat Block and Patch for Bonding Overcoat)
9.8 to 14	R7	SERV 52 XL	50% Alumina-Chrome Fired Brick (2nd Overcoat Layer)

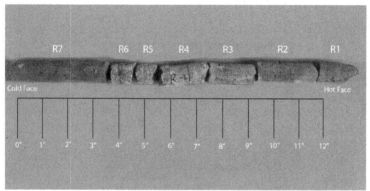

Figure 18. Photo of the core-drilled sample taken from the right sidewall.

On the right side where the core drill sample was taken, the original thickness of the JADECAST 50 was estimated at 10". Notice that the right core drill sample was 1" shorter than the left core drill sample. On the right side of the furnace the entire original SERV 30 DC sidewall block had been eroded away before the new working lining was cast.

Figure 19. Porosity Changes in Right Core Drill

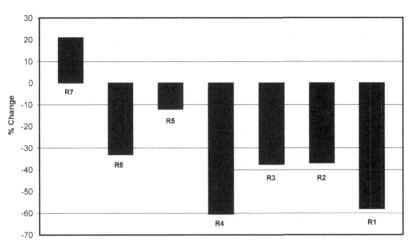

Figure 20. Amounts of Infiltrated Oxides in Lining

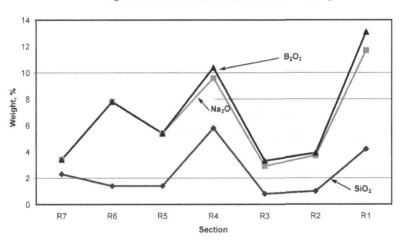

Figures 19 and 20 show features very similar to the left core drill sections. Section R 4 was definitely the previous working face because of its elevated boron oxide, soda and silica levels as well as its lower apparent porosity. The higher apparent porosity in R 7 was difficult to explain because this sample was infiltrated with molten glass. One would have expected the apparent porosity to be lower than typical for this refractory.

Microscopic Examination - Right Side of Furnace

Microscopic studies were also conducted on the samples from the right side. They showed very similar reactions as seen in the left side, but there was a difference between the depth of infiltration and the depth of significant reaction. The molten glass had infiltrated the right core drill sample to a depth of about 14", but significant reaction with the refractory was only seen to a depth of 1.5". Much less corrosion was seen in the other samples R 2 to R 7 although these sections contained more glass than the corresponding sections from the left core drill.

Scanning electron microscope examination of the used JADECAST 50 from the right core drill sample (R1, R2 and R3) showed glass penetration into all zones. The greatest glass penetration/disruption was present in R1; Samples R2 and R3 showed only minor glass penetration and disruption. The microtexture of R1 was characterized by fused chrome-alumina (shifted Eskolaite pattern) grains and a few remnant fused alumina-chrome (shifted Corundum pattern) grains in a matrix composed of an Na-Al-Si-O weakly crystalline phase and individual chromic oxide (Eskolaite) laths in an amorphous Na-Ca-Al-Si-O-rich glass (Figure 21). Samples R2 and R3 exhibited only minor disruption of the coarser fused alumina-chrome grains; no apparent alteration of the fused chrome-alumina grains were evident. The matrix primarily consisted of individual chromic oxide laths within a Na-K-Ca-Al-Si-O glass. R3 showed greater remnant porosity than R2; this suggested that less glass penetration and densification had occurred in R3 (Figures 22 and 23).

Figure 21   JADECAST 50 - Used (R1); 100x

1. Remnant Fused Alumina-Chrome Grain
2. Fused Chrome-Alumina Grain
3. Na-Al-Si-O Phase in Na-Ca-Al-Si-O Glass
4. Void/Epoxy

Figure 22   JADECAST 50 - Used (R2); 100x

1. Fused Alumina-Chrome Grain
2. Fused Chrome-Alumina Grain
3. Glass Penetrated Matrix
4. Void/Epoxy

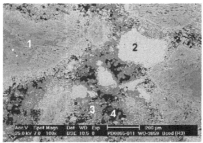

Figure 23   JADECAST 50 - Used (R3); 100x

1. Fused Alumina-Chrome Grain
2. Fused Chrome-Alumina Grain
3. Glass Penetrated Matrix
4. Void/Epoxy

SERV 30

Scanning electron microscope examination of the SERV 30 from the right core drilled sample (R4) showed minor sintering and penetration by an amorphous Na-Ca-Al-Si-O glass into the pore structure. A portion of the pore structure showed no glass penetration (Figure 24).

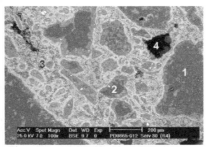

Figure 24   SERV 30 - Used (R4); 100x

1. Sintered Alumina Grain
2. Intermediate Sized Sintered Alumina Grain
3. Matrix w/Glass Penetration
4. Void/Epoxy

SERV 52 XL

Scanning electron microscope examination of the SERV 52 XL from the right core drilled sample (R7) showed only trace to minor glass penetration into the brick structure. No significant alteration was evident (Figure 25).

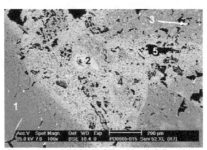

Figure 25   SERV 52 XL - Used (R7); 100x

1. Fused Chrome-Alumina Grain
2. Fused Chrome-Alumina Grain
3. Zirconia
4. Chromic Oxide-rich Matrix
5. Void/Epoxy

Additional properties of the cast working lining were examined to explain why this zone performs so well in service. Figure 26 shows the permeability of the castable over a range of temperatures. Notice that the permeability is extremely low, less than a half of a centidarcy, until the temperature reaches over 2000°F. At glass making temperatures, the permeability is still quite low at 6.3 centidarcies. This low value helps to limit the amount of molten glass that can penetrate into the working lining. For comparison, a typical fired alumina-chrome brick would have a permeability of 30 to 45 centidarcies.

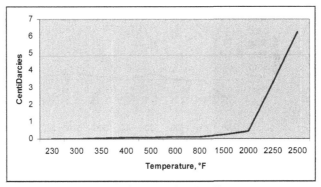

Figure 26. Permeability of JADECAST 50 vs Temperature.

Figure 27 shows the fine pore size of the working lining. After drying essentially all the pores are less than a micron. After heating to glass making temperatures about half of the pores are less than 10 microns in diameter. This fine size also helps to limit the amount of molten glass that can penetrate into the working lining.

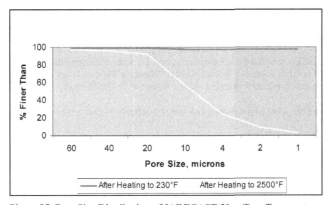

Figure 27. Pore Size Distribution of JADECAST 50 at Two Temperatures.

The strength of the installed working lining is also of importance. Figure 28 shows that after drying the castable is quite strong with a strength of about 1500 psi. This strength is maintained until about 800°F. At 2500°F, the strength has increased dramatically because of sintering of the chromia matrix and development of a ceramic bond.

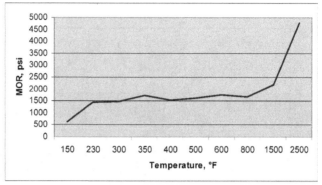

Figure 28. Strength of JADECAST 50 vs Temperature.

While the focus of this paper is to analyze the core-drilled samples from a single furnace, the conclusion would not be complete without updating and discussing the results to date and the benefits that wool glass producers have realized from the cast-in-place lining. As mentioned previously, 14 furnaces have been lined with the 50% chrome castable. The first ten furnaces were all cast between January 1997 and July of 2002. With respect to these 10 furnaces, one was shut down due to business conditions. Of the remaining furnaces, 3 were cast in 1998 and have all since been repaired. The cast lining in the three furnaces cast in 1998 had an average service life of 5.8 years. The first of these furnaces was shut down for repairs to the machines after only 4 years of service.

The remaining six of the original ten furnaces cast prior to the end of 2002 are all still in operation; one furnace was repaired with a rammable plastic in early 2005 after 5.5 years of service. The other furnaces are scheduled for repair in the next 1-2 years indicating that a predicted service life of 5-6 years and possibly more is a definite reality. Of course the actual service life is always dependent on operation parameters such a glass temperature, glass chemistry, and production rates.

Following are some of the proven and realized benefits of cast versus pre-fired block linings:

• Field experience has shown that the corrosion resistance is essentially the same for cast 50% alumina-chrome linings and for prefired 50% alumina-chrome blocks.
• Cast sidewall installation time can be less than block linings. Furnaces have been formed and cast in less than 36 hours.
• Casting allows use of existing sidewalls which serve as one of the form walls. This reduces demolition time and the quantity of used chrome material that has to be disposed.
• Casting eliminates the need to remove the cooling wind and furnace steel.
• Total project costs are lower. These include total disposal, labor and material costs.
• Faster relines are possible in the case of a premature drain or engineering changes. Castables can be manufactured and delivered faster than large block.

These successful casting installations have led to the next technology improvement: Shotkasting. The first installation of SHOTKAST 50% chrome-alumina refractory has been achieved. Total project costs were lower than casting and furnace downtime was reduced.

In conclusion, examination of the core-drilled samples after approximately four years of service indicated that complete penetration of glass into the new working lining had occurred. More glass had penetrated into the lining on the right side compared to the left side of the furnace, although the refractory matrix on the right side was not significantly altered. This difference in glass penetration and matrix alteration may be attributed to the difference in the thermal profiles of the working lining. When the new cast working lining was installed, the 30% chrome sidewall on the right side of the furnace where the core drilled sample was taken was completely gone. The right side of the furnace was essentially running on the overcoats which would have allowed for more glass penetration into the working lining, while the left side still retained approximately 3" of the original lining. This additional 3" of refractory on the left side would have provided more insulation and higher hot face temperatures on the left side resulting in increased corrosion of the matrix. Non-uniform furnace operating conditions may have further contributed to this difference in refractory matrix corrosion between the left and right sides of the furnace.

As discussed, corrosion of the chromia-rich matrix and the alumina-chrome fused grains in the working lining was evident. In spite of this, sufficient recrystallized chromia had formed during this corrosion process because the glass that forms quickly becomes saturated in chromia. This re-crystallization of chromia in combination with the low permeability and small pore sizes that are inherent to castables protected the working lining from rapid dissolution. This protection resulted in the working lining eroding only 3.5 to 4" during the 4 years of service, thus allowing for equivalent and even improved service compared to the conventional and pre-fired lining.

ACKNOWLEDGEMENTS

The authors wish to thank Owens-Corning for permission to core drill the walls of their furnace and to publish this paper. The authors are especially indebted to Rod Cook of Owens-Corning and to Leo Smathers of NARCO for their help in advancing the concept of cast sidewalls.

REFERENCES

1) D.R. Lankard, "Evolution of Monolithic Refractory Technology in the United States"; pp 46-66 in New Developments in Monolithic Refractories 13; presented at the International Symposium on New Developments in Monolithic Technologies, Pittsburgh, 1984.

2) E.A. Thomas and D.G. Patel, "The Application of Bonded Alumina-Chrome Refractories in the Glass Industry", Interceram (Special issue) (1986).

3) R. Cook, W. Fausey, M. Wheeler, D. Patel, L. Smathers, "Casting of a Chrome-Alumina Monolithic Lining for Melting Insulation Fiber Glass in a Cold-Top Electric Melter," presented at the 63rd Conference on Glass Problems, 2002.

STRUCTURE, MICROSTRUCTURE AND REFRACTORY PERFORMANCE

Nigel Longshaw
CERAM
Staffordshire, UK

ABSTRACT

The importance of modelling as a tool in design and development of refractory materials are considered. Both bulk refractory structures as well as refractory microstructures are considered. Theory on micromechanics of composites is applied to calculate effective material properties of microstructures. It is shown that microstructure modelling can be considered as part of the simulation driven material development process.

INTRODUCTION

Refractory and heat-insulating materials have the objective to manage and control high temperature processes economically. Furthermore, refractory materials help protect the environment by ensuring that high temperature processes do not have a harmful impact on our environment. Refractory materials can be stressed in the following ways:

- thermally by temperatures and thermal shock,
- chemically by gasses, liquids , melts, slags and
- mechanically by pressure, tensile force, friction and/or impact

Refractories are most always subjected to a combination of the above-mentioned stress factors. Consequently, the selection of appropriate refractory materials must take various stress factors into consideration. This is also true when developing refractory materials.

The role of microstructure modelling in the process of Simulation Driven Material Development will be presented in the following paper, the procedure of obtaining real microstructure images for modelling and the calculation of effective material properties of the microstructure and determine unknown material properties through back calculation of physical test results. The basic theory behind micromechanics and it application in computational modelling are presented.

It is widely accepted that refractories can be a key contributor to business profitability in the glassmaking industry. Refractory structures are critical in order to maintain functional and process stability. Whilst refractories can account for less than 3 % of the total installation cost of the process structure they can account for over 97 % of the structure and process functionality. It is vital that the refractory structures maintain themselves and do not under perform [1].

Refractory manufacturers are turning towards the aids of computational modelling in material development for the benefits that it holds: the reduction in capitol and operating costs associated with refractories, improvement on impact of design of high temperature equipment and operating efficiency and assist in the development and exploitation of new materials. Refractory manufacturers had raised concerns on the long time and excessive costs it takes to produce a new refractory material tailored for a specific application.

Historically, the development of refractory products has been based on industrial experience and empirical development. Such development methodology has led to large numbers of grades or compositions tailored towards specific applications. Finite element analysis (FEA) capability and computer power has increased dramatically over the last decade and it is now possible to investigate in detail the influence of different phases and refractory grains within the product. It is now possible to evaluate the effect of different micro & macro structures, for example predicting their resistance towards a range of industrial thermal environments. This can range from the performance of investment casting shells to crucibles to large refractory blocks for the glass industry.

Researchers [2] have reported that historically, the knowledge due to practise has permitted the designer to improve the materials i.e., their mechanical performance and consequently the life-time of refractory structures. However, by taking advantage of experimental methods (e.g., chemical and physical analyses) and multi-scale analyses, it is expected that further improvements of the performances of the ceramics can be achieved. However, microstructure modelling will still remain inevitably difficult due composition and manufacturing process. Therefore, one approach in modelling heterogeneous microstructures is making it a homogeneous material.

THE IDEA OF HOMOGENIZATION

Continuum mechanics deal with ideal homogeneous materials. Its aim is to describe their response to external loads using appropriate constitutive equations, thus, macroscopic experiments without microstructural considerations. The aim of micromechanics of heterogeneous materials is to derive their effective properties from the knowledge of the constitutive laws and spatial distribution of their components. Homogenization methods have been developed for this purpose.

Porous ceramics can be considered as a special case of multiphase mixtures, composites, or more generally, materials with microstructure. In micromechanics an effective stiffness tensor Ee can be defined via the linear constitutive equation in

Equation 1 where the angular brackets denote volume averages. In principle, the effective stiffness tensor can be predicted exactly when the properties of the constituent phases and all the details of the microstructure are known. In practice this is of course not the case [3].

$$\langle \sigma \rangle = E_e \langle \varepsilon \rangle \hspace{4cm} \text{Equation 1}$$

The idea of homogenization is to take a cube out of a linear heterogeneous medium of a micro-scale size and apply a load to this cube along its x axis. This cube will be referred to as the Representative Volume Element (RVE) and is explained later in this report. The stress component can then be taken as the load divided by the area and the strain as the extension divided by the original length. These calculations are obvious for a homogeneous RVE but not for heterogeneity results. Calculated experimental results of the stress and strain represent averages of the actual forces and displacements in the cube, also called volume averages (to be denoted by overbar). These calculated averages are mathematically represented in Equation 2 and Equation 3. This is the basic idea of homogenization of a heterogeneous medium (see Figure 1) and can be applied along any axis of the material. In turn, the Elastic Modulus can be calculated with Equation 4. V denotes the volume (RVE) under investigation.

$$\overline{\varepsilon_{11}} = \frac{1}{V}\int_{V}\varepsilon_{11}(x)dx \qquad\qquad \text{Equation 2}$$

$$\overline{\sigma_{11}} = \frac{1}{V}\int_{V}\sigma_{11}(x)dx \qquad\qquad \text{Equation 3}$$

$$E_{11}^{V} = \frac{\overline{\varepsilon_{11}}}{\overline{\sigma_{11}}} \qquad\qquad \text{Equation 4}$$

Thermal conductivity, k, can be calculated with Fourier's law of heat conduction (Equation 5) where the heat flux of a homogenous material is proportional to the negative of the local temperature gradient. Applying a temperature difference across a known microstructure distance enables to calculate the temperature gradient across the microstructure. Determining the spatial average directional heat flux make it possible to calculate the thermal conductivity of the microstructure through Equation 5 [5].

$$q = -k\frac{dT}{dx} \qquad\qquad \text{Equation 5}$$

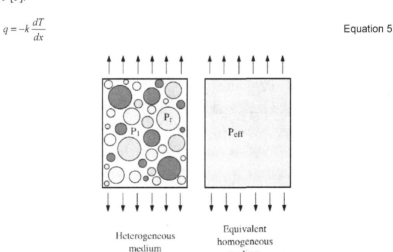

Figure 1: Illustration of the homogenization of a heterogeneous material.

Schmitt et al. [2] used micromechanics to highlight the role of each microstructural component of a refractory subjected to severe thermal loading. A generalized self-consistent scheme (GSCS) was used to obtain the homogenized macroscopic behaviour of the refractory. Strength tests were performed on two-component specimens (alumina-carbon) and through an inverse analysis, the properties of the components that were too difficult to measured directly, were obtained. The micromechanical model was validated through experiments. This procedure illustrated the advantage of using micromechanical modelling to determine material properties of the constituents.

Markov [3] employed two natural procedures to illustrate the basic ideas of homogenization of a composite material. These procedures are connected with the heterogeneous medium and its internal structure, independently of the specific choice of the sub-volumes. From these two procedures, the Ensemble Averaging was applied in this project.

ENSEMBLE AVERAGING

Perform a great number, N, of virtual experiments on differently centred cubes, identical in size and orientation, to determine each cube's material behaviour $E_{11}{}^{V1}$, $E_{11}{}^{V2}$, etc. These cubes are also referred to as the RVE. From these numerical calculations the average material property for all N samples can be calculated with Equation 6.

$$E_{11}^{*} = \frac{1}{N}\left(E_{11}^{V^1} + E_{11}^{V^{11}} + ...\right)$$ Equation 6

The meaning of this procedure is to obtain the expected gross behaviour of a heterogeneous medium through physical testing (i.e. three-point bend test). Equation 6 deals with the average reaction of a whole ensemble of specimens of identical shapes and sizes and applies identical external influence. The ensemble averaging is one of the basic notions in the theory of heterogeneous media of random constitution. See Figure 2 for illustration of this technique.

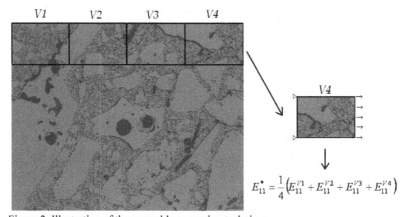

Figure 2: Illustration of the ensemble averaging technique.

THE REPRESENTATIVE VOLUME ELEMENT

A basic notion in micromechanics is the representative volume element (RVE). The block elements shown in Figure 2 are RVE's. This is a volume that is small enough from a macroscopic view so that it can be presented as a point in the heterogeneous medium, but large enough from a microscope view so that it will include a large number of inhomogenities and so will be a true representative of the microstructure. It is possible to describe representative volume elements which are small macroscopically and which dimensions define the mini-scale. This RVE consists of an amount of much smaller inhomogenities (inclusions, monocrystals, etc.) whose dimensions a micro-scale. The above mentioned principle is called the *MMM* principle and is described by the inequalities in Equation 7.

MICRO << *MINI* << *MACRO*   Equation 7

The position of a typical RVE is described by its coordinate X. Therefore, all continuum quantities like temperature, flux, stress, strain fields, etc., are then functions of X. The physical quantities of interest on the micro-level, i.e. within the RVE, depend also on the local coordinate x, e.g. the temperature $T = T(X,x)$ varies both on macro- and micro-scale levels. The connection between macro and micro quantities is supplied by the volume averaging, with respect to the micro-coordinate x where V is the RVE attached to macro point X (Equation 8). Since the point X spans the whole body, quantities like $\theta(X)$ in Equation 8 are called moving averages and the bar in the left hand side of the equation denotes the volume average across the RVE. Such averages play a central role in the elementary theory of effective properties.

$$\overline{\theta(X)} = \frac{1}{V} \int_V \theta(X,x).dx$$ Equation 8

In most cases, we will assume that the medium is statistically isotropic. This means that the macroscopic properties under study are independent of direction.

INVERSE IDENTIFICATION OF MATERIAL PROPERTIES

Due to the variability of elastic properties of ceramics, an inverse identification method is used to determine the material properties of unknown constituents in a heterogeneous/composite material.

Schmitt et al. [0] based this inverse identification model on an appropriate micromechanical model and the knowledge of one constituent. The measurement of the global property of the material through physical testing is an input in the calculation of the unknown characteristic. Material properties are known for most aggregate constituents and can be obtained from literature. The properties of the matrix is usually unknown and difficult (if not impossible) to determine though physical testing. The properties of the matrix are also dependent on other material and manufacturing features such as the packing density of the aggregate or the hydration of the mixture.

The Young's modulus of alumina aggregate and resin mix was determined through physical testing for different epoxy resin and aggregate volume content. The generalized self-consistent scheme (GSCS) was then used to determine the E value of the alumina aggregate. The GSCS was also used to determine the properties of the refractory binder.

This proposed homogenization procedure gave good predictions with refractory ceramics since the appropriate information on the microstructures was introduced. An inverse identification can be carried out to catch the local properties and to define mean properties for phases which are not easy to identify otherwise.

The GSCS method was not applied in this project, but the main principles in obtaining unknown material properties from this method, was used in the microstructure modeling.

MODELLING OF ALUMINA MICROSTRUCTURE

Applying the basic principles in micromechanics, it is possible to determine effective material properties of microstructures by applying virtual tests in computational modeling. By determining material properties such as elastic modulus and thermal conductivity, it becomes possible to characterize a material and evaluates its performance through modeling in an application, such as refractory lining [6], before the material is made. The procedure of microstructure modeling within the simulation driven material development process is presented in this section with two examples on different seized RVE's and the challenges in modeling these microstructural samples.

The basic steps in microstructure modeling can be summarized in   Figure 3 and pulls together the concept of Simulation Driven Material Development. The first step is to obtain microstructural images of the material. The digitization of the images into a format suitable for import into the finite element software proved to be an important step in the modeling process. Parallel with that goes the physical testing of the material for the material properties in question: the idea is to determine the unknown material properties of the microstructure constituents through inverse property calculation. In this case, the material properties of the alumina matrix were determined through modeling. With all unknown material properties determined, material microstructure could be evaluated in a virtual environment (Sensitivity Analysis).

DESCRIPTION OF REFRACTORY MATERIAL

A 99% alumina (Al_2O_3) refractory with low impurity content was used in this project. It is based on fused alumina and hydraulically cast using a reactive bond. The refractory has a wide variety of applications which are allied to its high degree of purity, temperature resistance and volume stability. Typical applications include ammonia reformers, high temperature kilns, sub-channel paving and refractory buffer courses.

The Al_2O_3 aggregates provide high stiffness and high strength at high temperatures. Fused cast alumina products usually have lower apparent porosity and thus more resistant to melt attack and are used in the glass industry, for instance, fused cast bricks containing chromia are installed in furnaces for glass fibers and borosilicate glass [4]. The matrix of the microstructure consists out of fine reactive alumina particles.

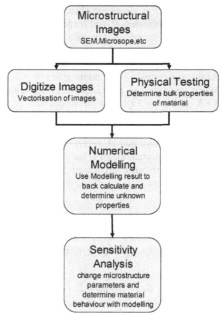

Figure 3: Flowchart of microstructure modelling procedure.

MICROSCOPY

Microstructure modeling can either use idealized microstructures made up of primitive geometries or real microstructures that can be obtained through microscopy. Building a microstructures using primitive geometries such as spheres, discs or triangular pyramids, allows for the creation of 3D structures with an interconnecting medium (matrix). Pores can also be introduced into the structure. This is however then an idealized microstructure and for the purpose of this project, it was decided to use real microstructures.

The alumina microstructure was investigated with a scanning electron microscope (SEM). The microstructure grains have clear boundaries in the images and the very small particles are well defined in the image. There is also a clear contrast between the constituents and the voids. A clear contrast is crucial for successful image digitization. Good contrast images can also be obtained through a polarized light microscope, but the reflective light microscopes has difficulty detecting the small particles within the matrix. Due to this effect, it becomes difficult to distinguish between the aggregate, matrix and voids/pores.

IMAGE DIGITIZATION

One of the key aspects in microstructural modelling is creating a real geometry and meshing of a microstructure: from SEM to FEM. This problem is solved by digitization of the picture file obtained through SEM. A raster image is made up from pixels, like the picture obtained from a scanner, or the screen image on a computer monitor. A vector image is stored as geometric objects, such as lines and arcs, which are drawn between specific coordinates. If you magnify a vector image you see the lines more accurately, and the line edges stay smooth. A raster-to-vector converter translates a raster image into the corresponding vector image.

Utilising raster-to-vector image translation, a microstructure image can be vectorised. The vectorised image can be opened in a CAD package and saved in format suitable for import in the FEA software. Figure 4 illustrates the process for an image vectorisation with an alumina-zirconia polycrystalline microstructure.

It was found that although the image quality (pixel count) is important for successful image vectorisation, the most important factor is the contrast between different grains. Figure 5 illustrates an alumina microstructure image that was used in the modelling

PHYSICAL TESTING

On the structural side of the model, three-point bend tests, apparent porosity and density of the sample were determined through the physical testing. These results were used as comparison with the numerical results and to verify the material behaviour in the modelling.

These results were also used as input parameters to the model and as verification parameters for the numerical results. The measured apparent porosity was used to compare with the validity of the sample of microstructure used in the modelling. The maximum load, maximum stress and modulus of elasticity were compared with the equivalent numerical results. The test results are given in Table 1.

Table 1: Material properties for high alumina samples obtained through testing.

Yield Displacement [mm]	0.0496
Yield Strain [%]	0.12
Max Load [N]	1678.42
Max Stress [MPa]	12.59
Modulus of Elasticity [GPa]	13.15
Apparent Porosity [%]	20.36
Bulk Density [g/cm³]	3.14

NUMERICAL MODELLING

Stress analysis was performed on each microstructure where bending and tensile conditions were simulated. The material properties for the alumina aggregate was assumed to be known and taken from literature and the bulk properties were determined through physical testing. The unknown material properties in the microstructure were the Elastic Modulus of the matrix. The matrix mainly consisted out of very fine alumina particles and was assumed to be an alumina body with high porosity. It was suggested that the change in elasticity with porosity is linear, therefore it was assumed that elastic modulus of the matrix was equal to the elastic modulus of the aggregate multiplied with a material factor. The material factor (MF) was changed until the numerical results were similar to the test results.

$$E_{matrix} = MF \times E_{aggregate} \qquad \text{Equation 9}$$

A bending condition was simulated by fixing one point on the side of the microstructure and applying a shear load on the opposite side. This loading condition created stress singularities in the model and incorrect results. The bending condition was done to perform a three-point bend test, but a three-point bend test is used in practise due to the difficulty of performing tensile tests on ceramic materials. This is not however the case in numerical modelling.

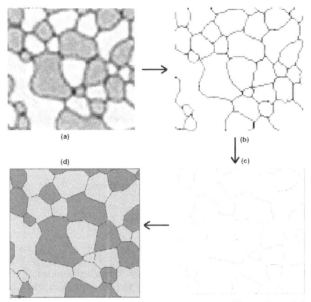

Figure 4: Diagram of digitization of microstructure picture file to FE model.

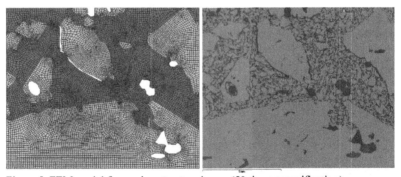

Figure 5: FEM model from microstructure image (50 times magnification).

Uni-axial tensile loading conditions were applied to a microstructure in the vertical and horizontal directions, thus, determining the Elastic Modulus of the microstructure along the two principle directions: x – and y axis. The spatial average stress and strain was determined through the modelling and used to calculate the elastic modulus (Equation 4). The modelling was done on microstructure images of 50x and 250x magnifications.

The material properties specified in the modelling for the alumina aggregate is specified in **Table 2**.

Table 2: Linear elastic material properties for Alumina

Material Property	
Elastic Modulus [GPa]	380
Poison's Ratio	0.26

1 X 250 MAGNIFICATION

The microstructure in Figure 6 is approximately 400 x 500 μm. No significant sized porosity or voids existed in the image and model consisted just out of two materials. Perfect bonding was assumed between the aggregate and matrix.

The areas of the microstructure also needs to be considered when a microstructure sample is chosen to see if it is representative of the overall bulk composition. Table 3 shows the value for the aggregate and particles areas and also the relative sizes, which is close to 50/50.

Table 3: Area representation of the 1 x 250 microstructure

Area	[mm²]	%
Aggregate	0.08395	42.9
particles	0.1116	57.1
Total	0.19555	100

Figure 7 shows the load vs. displacement results for the tensile loading conditions in the horizontal and vertical conditions. Material factors were calculated to be MF = 5 and 2.5% for the horizontal and vertical directions respectively, which convert to 19 and 9.5 GPa for E_{xx} and E_{yy} (see Figure 7). This meant the elasticity of the microstructure in the x – direction is twice that of in the y – direction. It was expected in the modelling that the higher the magnification of the microstructure, the more dependent the results will become on the aggregate size, grain boundary orientation and aggregate/matrix composition. The bulk properties of the material were assumed to be isotropic, but the microstructure modelling clearly indicated that this is not the case at micro-scale level.

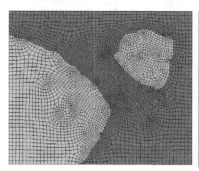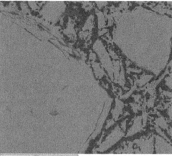

Figure 6: Alumina microstructure 1 x 250 magnification.

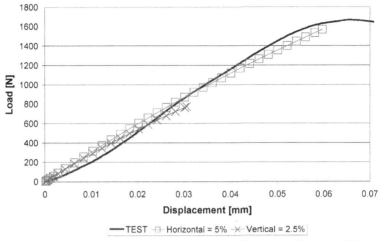

Figure 7: Load vs. displacement results for tensile loading in modelling: 1 x 250.

1 X 50 MAGNIFICATION

It was generally accepted that the bigger the microstructure sample, the less dependent it will become on the microstructure feature. This was not however the case for the microstructure sample used for the 1 x 50 magnification (Figure 8), as will be explained in below.

Table 4 tabulated the constituent composition of the microstructure in Figure 8. Not only is this microstructure sample much larger than the 1x 250 microstructure, but this sample also includes porosity/voids.

The results for this sample show large difference in material strength in the different directions. This has been attributed to the large aggregate at the bottom of the sample that spanned over the length of the sample.

Table 4 shows the area sizes and the percentage each component covered. The sample consists mainly out aggregate, 55.02% and particles, 42.37%. The porosity account for only 2.61% which does not represent a 20% apparent porosity of the bulk material. Not all pores or voids in the microstructure sample had been included and the model will require geometrical refinement.

In this sample there was a large difference in the values for the material factor between the two loading directions for tensile loading: <0.001% and 1.3% for the horizontal and vertical directions respectively. A low value for the MF indicates that most of the load in the sample is carried by the aggregate. Thus, the large aggregate at the bottom of Figure 8 dominated the stiffness of the microstructure sample, resulting in highly anisotropic results. It was therefore suggested that when choosing microstructure samples, to avoid samples where one aggregate spans across the entire dimension of the sample.

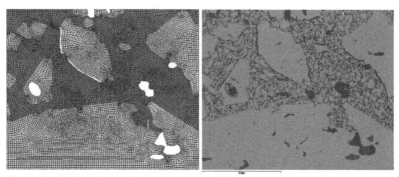

Figure 8: Alumina microstructure 1 x 50 magnification.

Table 4: Area representation of the 1 x 50 microstructure

Area	[mm²]	[%]
Aggregate	2.6718	55.02
Particles	2.0577	42.37
Porosity/voids	0.12654	2.61
Total	4.85604	100

CONCLUSION

Refractory materials and structures are vital to the performance of the glass production process. Modelling enables development of new materials reducing the need to make numerous material samples and testing of it until the right material has been developed. The range of structures, materials and environments that can be considered is vastly increased by the use of models, making modelling a value added rather than cost added part of the design process.

REFERENCES

1. George, S.D., Cronje, M., Farn, S.M., 2006, Modelling High Temperature Refractory Structures for the Glass Furnace, Advances in Science and Technology, Vol. 45, pp. 2308 – 2315.

2. Schmitt, N., Burr, A., Berthaud, Y., Poirier, J., 2002, Micromechanics Applied to the Thermal Shock Behaviour of Refractory Ceramics, Mechanics of Materials, vol 34, pp 725 – 747.

3. Markov, K.Z., 1999, Elementary Micromechanics in Heterogeneous Media, Heterogeneous Materials – Micromechanics Modelling Methods and Simulations, edited by K.Z. Markov and L. Preziosi, Birkhauser, Boston. 1999, pp 1 – 162.

4. Deutsche Gesellschaft Feuerfest- und Schornsteinbau e.V., 2005, Refractory Engineering: Material, Design and Construction, 2nd Revised and Updated Edition, English Edition, Vulkan-Verlag GmbH, Germany.

5. Mills, A.F., 1995, Heat and Mass Transfer, Richard D. Irwin, Inc. Chicago, USA.

6. Poirier, J., 2003, Thermomechanical Simulations of Refractory Linings: An Overview, Refractories Applications and News, Vol.8, No.6, pp16 – 22

Author Index